THE
PRACTICAL
ORNITHOLOGIST

THE
PRACTICAL
ORNITHOLOGIST

John Gooders
Scott Weidensaul, Editor

All pictures supplied by
Bruce Coleman

A FIRESIDE BOOK
Published by Simon & Schuster Inc.
New York   London   Toronto   Sydney   Tokyo

A QUARTO BOOK

Simon and Schuster/Fireside
Simon & Schuster Building
Rockefeller Center
1230 Avenue of the Americas
New York, New York 10020

First published in Great Britain in 1989 by Quarto Publishing plc

Designed and produced by Quarto Publishing plc, The Old Brewery,
6 Blundell Street, London N7 9BH

Designer: Hazel Edington
Picture Manager: Joanna Wiese
Assistant Art Director: Chloë Alexander

Art Director: Moira Clinch
Editorial Director: Jeremy Harwood

Typeset by Ampersand Typesetting Ltd.
Manufactured in Hong Kong by Regent Publishing Services Ltd
Printed by Leefung-Asco Printers Ltd, Hong Kong

1 3 5 7 9 10 8 6 4 2
1 3 5 7 9 10 8 6 4 2 Pbk.

Library of Congress Catalog Card Number: 89-21902

ISBN: 0-671-69302-6
0-671-69301-8 Pbk.

CONTENTS

6 .... FOREWORD

BIRDS AROUND US

8 .... INTRODUCTION

10 .... YARD BIRDS

12 .... FEEDING BIRDS

16 .... BIRD GARDENING

HOW BIRDS WORK

18 .... DESIGNED FOR FLIGHT

22 .... FEATHERS

24 .... ANATOMY

28 .... THE BIRD'S YEAR

30 .... MIGRATION

34 .... SONGS AND CALLS

38 .... BIRD NAMES

BIRD BEHAVIOR

40 .... TERRITORY

44 .... NESTS

48 .... EGGS

50 .... INCUBATION AND REARING THE YOUNG

54 .... LIFESPANS

LOOKING OUTWARD

56 .... BINOCULARS

58 .... TELESCOPES AND TRIPODS

60 .... THE COMPLETE BIRDER

62 .... IDENTIFICATION

64 .... FIELD GUIDES

66 .... FIELD NOTEBOOK

68 .... PHOTOGRAPHY

70 .... RECORDING

72 .... LISTING

74 .... THE BIRDER'S YEAR

URBAN BIRDING

76 .... CITIES

78 .... SUBURBIA

80 .... DUMPS

82 .... SEWAGE PONDS

84 .... OASES IN CONCRETE

FRESH WATER

86 .... MOUNTAIN STREAMS

88 .... LOWLAND RIVERS

90 .... PONDS AND LAKES

94 .... BOGS

96 .... SWAMPS AND LAKES

SALT WATER

100 .... CLIFF COLONIES

102 .... DUNES AND ROCK BEACHES

104 .... BETWEEN THE TIDES

106 .... ESTUARIES

110 .... THE OPEN OCEAN

CULTIVATION

112 .... FARMS

114 .... GRASSLANDS

116 .... BRUSHY AREAS

FORESTS

118 .... DECIDUOUS

120 .... CONIFEROUS

122 .... MIXED FORESTS

124 .... SUBTROPICAL

OPEN COUNTRY

126 .... SEMI-DESERT SCRUBLAND

128 .... DESERT

130 .... SAGEBRUSH FLATS

132 .... ARCTIC TUNDRA

HILLS AND MOUNTAINS

134 .... HIGH PEAKS

136 .... ALPINE TUNDRA

138 .... ALPINE FORESTS

FROM BIRDER TO ORNITHOLOGIST

140 .... ETHOLOGY

142 .... CENSUSES

144 .... POPULATIONS

146 .... BANDING

150 .... MIGRATION STUDIES

152 .... CONSERVATION

154 .... CONSERVATION PROJECTS

156 .... GLOSSARY

157 .... INDEX

160 .... ACKNOWLEDGMENTS

FOREWORD

Watching birds comes easily. Primitive man watched birds for his own reasons but, while he sought to understand them to catch them for the pot, he could not have failed to find them fascinating in their own right. He noted their regular appearances and disappearances, their strange antics in spring, their marvellous songs which doubtless he tried to imitate and, above all, he wondered at their powers of flight.

Not surprisingly birds figure largely in primitive art, from the caves at Lascaux and Altamira in Europe, to the rock paintings of Bushmen in the Kalahari and the Aborigines in Australia. And, in the few primitive cultures that survive in our shrinking world, people still refer to birds by their culinary qualities. The people of New Guinea offer a fascinating example of primitive man's relationship with birds. For while birds have been given names such as "Large Eating Bird" and "Small Non-eating Bird," their beautiful feathers are also widely used in symbolic decoration.

The appeal of birding

Today there is little obvious wildlife around us except birds. That perhaps explains why watching them has become so important to so many people during the 20th century. Certainly, the sport of birding, with its emphasis on the pursuit of birds, can be seen as the sublimation of the hunting instinct. Yet, just like our predecessors, we seek not only to locate and see birds in their natural environment, but also to understand them and their lives.

It is the aim of this book to lead from this fascination to a deeper understanding of the only form of wildlife that has been able to survive in number and variety alongside us. It is a celebration of birds and is written in the hope that the reader will develop his or her own fascination for birds and become a more informed and skillful watcher as a result.

The great blue heron is widely distributed through North America, though it breeds only erratically in many areas. It avoids the worst of the winters in the interior by migrating to the more temperate coastlines.

BIRDS AROUND US: INTRODUCTION

Birds vary enormously in size, shape, color, length of life, habitat, population, food, hunting methods, and so on. Since they first appeared about 150 million years ago in the Jurassic period, they have managed to occupy virtually every part of the planet Earth. Even in the most hostile of environments, that is the regions surrounding the two poles, they manage to utilize some elements of the landscape to their advantage. Indeed some species actually breed within a few hundred miles of the South Pole, including the remarkable emperor penguin which incubates its single egg on its feet on the frozen sea in the depths of the Antarctic winter.

Birds may range in size from huge eagles to tiny wrens yet, I would venture, no one would have any difficulty in identifying any of the world's 8,600 plus species as birds. One of the most important factors in this apparent uniformity of appearance is the structural requirements of flight. Even the

Left The emperor penguin places its single egg on its feet, where it is incubated by a special fold of skin.

Above Ostriches have evolved from flying birds to running birds that avoid danger on foot rather than in the air.

huge flightless birds that inhabit the southern continents are descended from birds that could fly at one time in the distant past. Their loss of flight was a simple adaption to a particular environment, whether it was to the pampas of South America, like the rheas, or the forests of Australia, like the cassowary. The flightless penguins have swapped flying through the air for "flying" through the water. But for most birds flying is an integral part of their lifestyle.

Flight does, of course, make birds the most mobile of all animals enabling them to take advantage of distant food sources and make lengthy journeys to enjoy seasonal abundance. But all birds, be they migrants or residents, rely on

lucky to live to be two or three years old.

Such different sizes and routines are typical of the variety of ways in which birds live. Yet these are two relatively common species found in the northern part of our planet. Elsewhere there are birds that live in the most remarkable of ways. There are birds that construct incubators instead of nests; birds that feed on parasites that in turn live on large mammals. There are birds that use echo-location just like bats, and birds that dive 600 feet to the seabed to feed. There are birds that behave like locusts, devouring everything in their path, and are numbered in the billions, while others are so rare that they may only be seen once in a generation. There are birds that spend their lives at sea, circling the globe for years on end, and others that fly over the world's highest mountains.

Left Golden eagles have a home range covering between 2000 and 7000 hectares, which may contain up to 12 different nest sites.

Below The tiny house wren needs no more than a crevice in a tree to hold its nest, yet it may lay 10 or more eggs and rear more than one brood each year.

their powers of flight to escape predators and it is this ability that has enabled them to survive while slower, land-based creatures have succumbed to human predation. Indeed, it was not until the invention of the shotgun that humans made any serious inroads into the populations of most wild birds.

Bird specializations

Our fascination with birds is, then, due partly to their ability to live alongside us and partly to their aerial mastery and mobility which enables them to occupy virtually every part of the planet. Along the way they have had to specialize and this, in turn, has led to the development of a huge variety of different shapes and sizes. Golden eagles soar over remote mountain ranges, eyes alert for ptarmigan or hare. They hold territories that are measured in square miles rather than square yards. In a normal season they will lay two eggs, but rear only one youngster, which will take three or four years to reach maturity. The tiny wren, on the other hand, skulks deep in tangled thickets and lays five to eight, sometimes as many as 16 eggs. Its territory is small, not much larger than a decent-sized back yard, and its young are unlikely to survive more than a few months. It breeds after one year and is

Such is the diversity of birds. It is perhaps not surprising that we find them so intriguing.

YARD BIRDS

Though the birds of the world are so variable most of us start our interest in them in our own back yard. We watch them come boldly to the feeders and bird tables in winter. We note the aggressiveness of one, the quiet dominance of another and a general pecking order that ensures that serious fighting is avoided. We see pairs beginning to form as the harsh weather of winter gives way to the lengthening days of spring, and we puzzle to tell male from female.

With some species like the American goldfinch or cardinal it is easy to tell the sexes apart – males are simply brighter than females – but with others we may have to search for smaller plumage differences. With the North American sparrows, for instance, most species are virtually impossible to sex accurately. An exception is the dark-eyed junco, a complex of four geographic races found across the continent. The male of the "slate-colored" form is a darker gray than his brownish mate, while the only difference between the sexes in the "Oregon" form of the Pacific Northwest is the intensity of contrast between the male's black hood and brown back.

Distinguishing characteristics
In the case of the American robin the female has a paler head than the male, but this again needs patience and a little experience to pick out accurately. The European robin (a member of a different bird family), however, is utterly impossible to sex on plumage characteristics and one must resort to behavioral features. In late winter and early spring, males and females will often visit feeders together. Sometimes one will adopt a submissive posture that involves lowering the red breast (the female) while the other boldly shows its breast to advantage (the male). As spring develops it is the male that does most of the singing, though both sexes sing in winter.

Gradually we get used to the birds that visit our gardens. We learn to name them quickly and accurately and become more expert at picking out the less frequent visitors. We probably keep a list, if only in the mind, of birds seen, so that a new

Left Cardinals are among the most colorful birds that the bird gardener can hope to attract.

Above Bird feeders attract a wide variety of different species, especially when there is snow on the ground. Here great and blue tits, chaffinch, greenfinch, and the more unusual brambling and hawfinch have come to feed.

Below Blackbirds, like this female, are regular visitors to suburban bird tables where a wide variety of food will be taken.

Right The house finch is a bird of western North America that has expanded its range in the east and is often seen at bird feeders.

or back yard. Already you have noted that different birds feed in different ways. Some come readily to bird feeders, others to hanging nut bags. Some can "unwrap" hard-coated seeds, while others can do no more than pick up small morsels. Some will never actually alight on the bird feeder but will find what they need on the ground, feeding on the fallen crumbs. The secret of good bird feeding is to know what to offer at what times to ensure the greatest range of species. But this principle applies equally to the whole field of bird gardening, for there is much more than food that can be offered to attract birds.

Water to drink and in which to bathe is almost as important as food and, in drier areas, may prove an even greater attraction. Providing nest sites will ensure that birds will spend the whole summer in the garden, and do not forget nest materials either. Vegetation, that is gardening with birds in mind, can offer food, shelter, roosting sites and, of course, nest sites to birds that will never accept an artificial alternative. Finally, if the garden is large enough, an overgrown pond can offer a home to different groups of birds as well as providing nesting materials and an abundant source of insect life to the more regular garden birds.

In today's world the creation of habitats has taken on a new importance. No longer do we need to preserve a particular marsh or meadow when we can recreate another nearby. The bird gardener can do his bit by understanding the needs of birds and by thinking out an active management of what he can, with a little effort, provide.

garden bird becomes a source of great interest. The appeal of listing the birds will quickly lead to extending the garden skyward to include those birds, like the gulls and pigeons, which seldom if ever land, but which use the garden's "air space," At this point one should beware, for it is a short step from noting birds flying over a garden to venturing out to see what birds are present in the local park or surrounding countryside. I say "beware," because becoming a birder can seriously damage other aspects of one's life. It can become a passion that is all but impossible to shake off.

The birder and garden birds

Long before you become a dedicated birder, there is still much, much more to do around the garden

FEEDING BIRDS

Inevitably the beginner bird watcher will want to feed the birds around the home in winter, and it is an easy step to visit a local garden center or hardware store for a simple feeder and a bag of seed. Hung from a tree branch or clothesline pole, the feeder will quickly attract (depending on habitat and the part of the continent) chickadees, titmice, pine siskins, jays, woodpeckers, and other species. Cardinals, juncos and mourning doves will appear to scavenge the seeds that have fallen to the ground.

Choosing a feeder

The number and variety of feeders offered for sale is staggering, but, unfortunately, not all are well-designed. There are several major categories. Vertical tube feeders are made of clear plastic, with perches and feeding portals along their length; filled with seed, they attract the smaller, more nimble species, such as chickadees. Shelf feeders may be as simple as a flat board nailed to a stump, or as elaborate as those made with storage hoppers that automatically spill more food onto the shelf as it is consumed. Suet – beef fat – is coveted by woodpeckers and other winter birds, and requires a special feeder – either a nylon mesh bag like those that onions are sold in, or a basket of galvanized hardware cloth, which can be hung from a tree.

There are endless variations – clear plastic feeders that affix to a window through the use of suction cups, coconut shell feeders for seed, suet logs with a dozen or more holes drilled to hold fat or peanut butter, even spring-loaded feeders designed to exclude squirrels and large birds. But birds care little for fancy feeders, and there is no reason to be extravagant. For most home feeding stations, a simple shelf feeder, a tube feeder, and a wire suet basket will fill all needs.

Suitable foods

The choice of food is probably more important than the feeder design. By far the most universally attractive food is sunflower seed, especially the

warbler

finch

wader

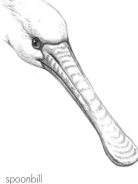

spoonbill

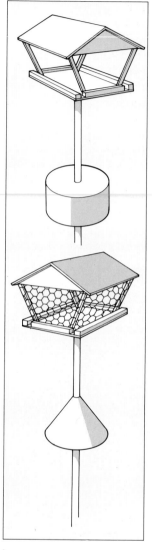

Above Bird tables provide feeding opportunities for birds that cannot cope with hanging food dispensers, thus widening the range of birds that will use the garden. The rim prevents food from falling off; a roof protects against rain; a guard deters mice and squirrels; and wire mesh enables small birds to feed while discouraging large ones.

Left Bills are adapted to different foods or different feeding methods. Warblers snap up insects; finches crack seeds; waders probe in mud; spoonbills sift through ooze.

small, all-black variety known as oil sunflower, which has more nutrition and less waste than the more common gray-striped variety. Everything from tiny redpolls to big evening grosbeaks relish sunflower seeds, which can be placed on an open shelf or used to fill a tube feeder. If possible, buy seed directly from a feed mill, which is more economical, or from a garden center that sells it in bulk. The least economical alternative is to purchase small bags of mixed seed from a grocery store – the per-pound price is very high, and the mixes contain a great quantity of "waste" seeds, like rape, which few birds eat.

While sunflower seeds are the best all-round choice for feeding, there are several other foods that particular birds enjoy. White or red proso millet is one of the best alternative foods, and is a clear favorite with mourning doves, juncos, and most sparrows; because these species prefer to

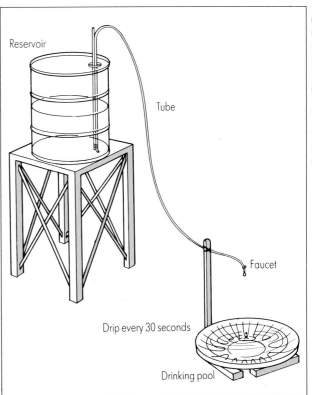

Above Water is essential to birds and many different species will come to drink and bathe. Moving water is particularly attractive to warblers and a regular drip system will bring in birds that are otherwise seldom seen in gardens.

feed directly on the ground, it is often a good idea to scatter the millet, rather than fill a feeder with it. Fine-cracked corn is attractive to many of the same birds, as well as ring-necked pheasants and bobwhite quail, should you be lucky enough to have them visiting your feeder. "Thistle" seed (actually niger), a tiny black seed imported from Africa and Asia, has a powerful appeal for goldfinches, siskins, and other small seed-eaters, but the cost can be exorbitant – especially since these same birds will happily eat oil sunflower.

Peanuts and peanut kernels are frequently recommended as bird food, but in actuality, few species eat them. Blue jays and nuthatches will hammer a peanut shell open, and titmice like cracked kernels, but avoid peanut hearts, which mostly attract starlings, a terrible pest at the feeder. By the same token, stale bread is also best avoided; its nutritional value is slight, and it will draw house sparrows in profusion.

Above A seed hopper like this will prove a regular feeding place for species such as these red-winged blackbirds in Florida. The large, pale-eyed birds are Brewer's blackbirds.

13

Feeder locations

Many beginning birders buy a feeder, fill it with food, and then wonder why no birds use it. Location is usually the reason. At the feeder, a bird is exposed to attack from cats, hawks, and other predators, so if the feeder is placed too far from cover, the birds may simply refuse to patronize it. This is especially true in new housing developments, where there are no large trees anywhere in the area. The ideal situation is to hang the feeder from the branches of a tree, with thick escape cover – like conifers or dense shubbery – close at hand. A discarded Christmas tree, wired to a stake to keep it upright, can provide a winter's worth of protection in an otherwise barren yard.

Left White-breasted nuthatches can be encouraged to come to feeding stations by forcing fat, nuts and other food into bark crevices or tree holes.

Above The blue jay is a common bird over most of North America west of the Rockies. It is a regular visitor to many suburban yards and comes readily to bird tables.

The sort of habitat will also dictate what kinds of birds come to your feeder. A feeding station in eastern woodlands will probably attract northern finches like grosbeaks, as well as black-capped or Carolina chickadees, downy woodpeckers, blue jays, and white-breasted nuthatches. A feeder in open farmland will lack most of the forest species, but will likely host such grassland and brushland birds as juncos, American tree sparrows, goldfinches, mourning doves, cardinals, and song sparrows. Suburban feeders will have a mix of both groups, along with house finches, mockingbirds, grackles and starlings.

Potential problems

A bird feeding station is not without its difficulties and problems. The most serious is also the least visible – a lack of cleanliness which can prove fatal to your bird guests. Feeders should be emptied and scrubbed regularly, especially in mild spring or summer weather, or during particularly wet periods. Remove seed that has become soaked – mold and fungus infestations may kill birds that feed upon it, and are a powerful argument for the use of covered feeders rather than exposed shelves. Take down any suet feeders when the weather turns mild, because the warm fat will turn rancid quickly, and furthermore will mat the feathers of the bird feeding on it, reducing their insulation abilities in rain.

In some places, rodents may be a problem, because mice and rats relish seeds as much as birds. Store bird food in metal canisters with tight-fitting lids, and put out no more food than your birds will clean up in a day. An expensive alternative is to feed them shelled sunflower seeds, which generate less waste. Aluminum sleeves on feeder poles will also keep rodents from climbing to the feeder itself.

Gray and red squirrels are among the biggest pests at a feeding station, hogging the food, chasing away the birds, and often damaging the feeders.

There are many baffles and other gadgets that are said to "squirrel-proof" feeders, but none is completely effective. For the best protection, place feeders on tall poles armed with metal sleeves, out of jumping range from trees and the house. Never underestimate a squirrel's acrobatics or ingenuity, however; they have been known to walk an electric line like a high-wire expert, then hang from their hind legs to drop onto the feeder below.

In wooded areas and suburbia, feeding stations in winter often attract hawks, especially the swift bird-eating species like sharp-shinned and Cooper's hawks. While it is traumatic to see a predator kill another animal just outside your living room window, please remember that the hawk, too, must eat to survive. By culling the slow and unsteady from the flocks, the hawk is actually doing the birds as a whole a service, especially in

preventing the spread of disease. Make sure there is escape cover near your feeding station, and try to enjoy the thrill of seeing nature at work, should a hawk begin patrolling your yard.

Above Evening grosbeaks are mainly winter visitors; feeding stations make a major contribution to their survival during the hard months of the year.

Left Resident throughout most of the eastern United States, the tufted titmouse is as frequent a visitor to bird feeders as other members of the chickadee tribe.

15

BIRD GARDENING

Nest boxes are freely available at most gardening centers, shops and hardware stores. They come in a variety of shapes, sizes, and designs, and the vast majority are useless. It seems obvious to me that nest boxes should be designed to appeal to birds rather than to people, yet most of those available through retail outlets work on exactly the reverse principle. Birds will occupy boxes that conform to the criteria that they apply to natural holes. They should be out of direct sunlight; that is, face northward in the northern hemisphere and southward in the southern hemisphere. They should be away from the prevailing winds; that is, face eastward over most of North America, though subject to local circumstances. They should be large enough to accommodate a brood of 10 or more virtually adult-sized birds; about 6in square floorspace, with an entrance hole sized to fit the particular bird you want to attract – $1^1/_4$in for house wrens, chickadees, titmice, and nuthatches, $1^1/_2$in for bluebirds, $2^1/_2$in for flickers. Such boxes can be made at home, purchased from national or local bird societies, or via mail order from advertisers in bird magazines. However, there is one other major requirement that applies to all nest boxes; they must be easy to open, for after use at the end of the breeding season boxes should be cleaned out ready for winter roosting and the next breeding season.

Varieties of next boxes

There are, of course, many other artificial nests than the standard "holed entrance" box. An open-fronted box will attract robins and phoebes. Larger boxes can offer a home to owls, provided they are placed high enough in a tree, or even kestrels if they are right at the very top. In fact any hole-nesting bird is a potential nest box user provided that the right shape and size are offered in the right position. Some bird gardeners have gone to enormous lengths to try to satisfy the needs of

Left Eastern Bluebirds are traditional nest box users that have suffered severely in competition with the introduced house sparrow which takes over their nest sites.

Above The purple martin would doubtless be a much rarer bird were it not for the provision of special communal nest boxes that act as tenements for these birds.

particular species. Realizing that the European willow tit always excavates its own nest hole in a rotten tree stump, one enthusiast filled an ordinary "holed" nest box with expanded polystyrene so that the birds could hack out their ideal home from this soft material. And it worked. Imagination, thought, and a good understanding of a particular

Nest box construction

Three easy stages in nest box construction. The use of good quality wood and securely glued joints makes for a weatherproof home for the inhabitants.

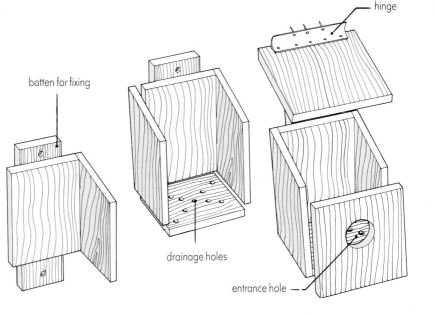

batten for fixing

hinge

drainage holes

entrance hole

bird's requirements can bring a great deal of satisfaction if the box offered is a success.

One of the most successful "good ideas" was the construction of multicelled nest boxes for purple martins. These birds formerly bred in hollow trees, crevices in cliffs, and old woodpecker holes, and still do so in remote areas. They have, however, taken readily to specially constructed boxes set on poles 15-20 feet high, offering customized apartments for 10 to 20 different pairs.

The European house martin builds a neat, gourd-shaped structure of mud beneath the eaves of town and village houses. They can be encouraged by placing a row of specially constructed replicas in similar positions. Barn swallows nest in barns and outbuildings and the erection of a short, narrow ledge can often encourage them to occupy an otherwise unsuitable site. Even sand martins (bank swallows) will take to an artificial bank with horizontal oval holes backed by a sand cliff.

The complete bird gardener will, however, not only be offering food, water, and places to nest, but will also attempt to supply everything else a bird needs. Nesting materials may be in short supply and the provision of a dispenser holding soft feathers, sheep's wool, and even lichen may attract a good variety of species. Offering fallen apples during the depths of winter will bring in thrushes

Above Best known for their imitations of a huge range of sounds – both natural and unnatural – mockingbirds are highly territorial in winter as well as summer.

that seldom visit bird tables.

Alternatives to nest boxes

Actually, there is no limit to what ingenious bird gardeners can do to bring birds into their domains. Planting suitable berry-bearing shrubs, such as hawthorn and cotoneaster, elderberry and cherry, are obvious if the back yard is large enough to accommodate them. Leaving piles of uncleared brushwood offers nest sites to birds.

HOW BIRDS WORK: DESIGNED FOR FLIGHT

One of the most important features of birds, shared by the vast majority of them, is the structural adaptations necessary for flight. The ability to fly makes them highly mobile but imposes strict limitations on their diversity. The heaviest flying birds weigh no more than 40 pounds, whereas those that have successfully dispensed with the need to take to the air may weigh several hundred pounds. There is a complex interrelationship between weight and power that effectively means that most birds, along with other flying animals, such as bats, are small and light.

Flightless birds

The ostrich has dispensed with flight by developing long powerful legs and antelope-like feet. It avoids danger not by flying but by running very fast. Other species, notably rails, have dispensed with flight because of the lack of ground predators on their island homes. Once established on a predator-free island, these already weak fliers quickly adapted to new circumstances in geographical isolation. They have no need of flight. Only during the last century or so have these endemic rails suffered a decline due to human introductions of alien predators such as rats and cats. The flightless cormorant of the Galapagos Islands was able to dispense with flight simply because of the immense richness of the seas around its island home, whereas the world's other cormorants need to fly long distances to the richest fishing grounds.

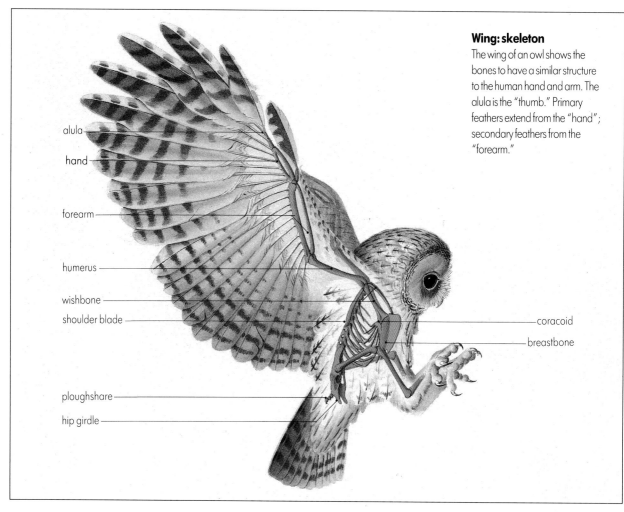

alula
hand
forearm
humerus
wishbone
shoulder blade
ploughshare
hip girdle
coracoid
breastbone

Wing: skeleton
The wing of an owl shows the bones to have a similar structure to the human hand and arm. The alula is the "thumb." Primary feathers extend from the "hand"; secondary feathers from the "forearm."

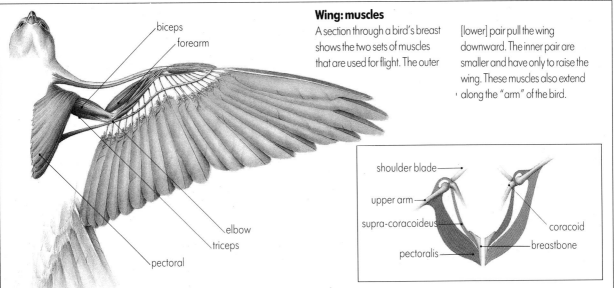

Wing: muscles

A section through a bird's breast shows the two sets of muscles that are used for flight. The outer [lower] pair pull the wing downward. The inner pair are smaller and have only to raise the wing. These muscles also extend along the "arm" of the bird.

biceps
forearm
elbow
triceps
pectoral

shoulder blade
upper arm
supra-coracoideus
pectoralis
coracoid
breastbone

Below The spread wing of a barn owl shows the elaborate structure that is required for flight. The small oval feather protruding from the bend of the wing is an anti-stall device called the alula that smooths the passage of air over the wing as the bird comes in to land.

The flightless cormorant propels itself underwater with its huge webbed, paddle-like feet, whereas those other marine fish eaters, the penguins and auks, have developed even further and swim beneath the surface with their wings. In the case of the penguins this process has been taken to the logical conclusion of flightlessness simply because of the safety of the areas they inhabit around the Antarctic continent and the southern seas. The northern auks, save for the extinct great auk, never developed so far. They are thus not the perfect underwater fishermen that the penguins have become, nor are they very good fliers. Their wings serve two distinct functions; they are a compromise.

Why birds fly

Flightlessness has, then, evolved among extra large birds that can outpace or out-fight their enemies; among birds that find their food without traveling huge distances at sea; and among birds that have colonized isolated islands free of predators. It follows that flying birds, that is the vast majority, need flight just for these purposes, that is, to escape enemies, to find food, and to travel long distances. Naturally there is a huge variation in the methods by which these needs are met. The average grouse, for example, uses the power of flight solely for protection and is incapable of sustained long-distance journeys. Gannets, on the other hand, have few natural predators and rely on their aerial powers to search for and obtain food. The swifts similarly capture their prey in the air, but also make lengthy migrational journeys to enjoy the long days of summer throughout the year.

Birds are adapted to flight in a number of ways, and their ability to fly has a profound effect on their lives.

Wing Shape

While all flying birds have a basic wing and feather structure in common, the actual shape of the wing varies enormously. Some birds are highly aerial, fast fliers like the swifts. Their wings are long and pointed and their streamlined body shape fits their mode of living. Vultures are also highly aerial, but unlike swifts, they do not feed in the air and thus do not require the speed and agility of the swifts. Their long, broad wings enable them to circle slowly overhead, ever watchful for signs of carrion or dying creatures, and their stalling speed is extremely low.

Hummingbirds have extremely long and attenuated wings that beat so fast they produce the audible hum which gives them their name. They are the complete aerial masters, capable of hovering before a flower to feed and even of flying backward. Their remarkable abilities stem, in part at least, from their unique ability to use both the downsweep and the upsweep of the wings to generate power.

Above Most of the world's birds can fly, but none fly as far as the Arctic tern, which covers 22,000 miles a year on migration alone. Just how far these birds fly in a year is a mystery, but it is certainly several times the mileage of the average car.

Feather functions

This contrast in wing shapes is best exhibited by birds that live totally different types of lives. Grouse, for example, are ground-dwelling land birds and escape predation by concealment rather than flight. When forced into the air, their rounded wings produce a sudden burst of acceleration and they then glide low over the ground before alighting again. In sharp contrast the wings of the albatrosses are long, narrow, gliding wings which support the bird with barely a flap. In this case propulsion is provided by the force of the wind and the upcurrents in the oceanic wastes that they inhabit.

While the feathers of the wings and tail provide lift and propulsion in flight, these are not their only functions. In some groups, such as waders, they are boldly marked with bars that serve as identification and contact marks in fast-flying flocks. Some species, notably the pennant-winged nightjar, even have specially extended feathers that protrude from the wing and are part of territorial and nuptial displays. While such appendages are unusual and would clearly seem to impede a bird in flight, tail streamers are comparatively commonplace. The tail of the resplendent quetzal of South America is three times as long as the body; and nearer home the skuas have all evolved long streamers in breeding plumage.

Body plumage does not only act as an insulator. Most birds are colored, and for a variety of reasons. Many are camouflaged so as to avoid predators. Others are boldly colored to attract mates or defend territories. Others have patches of bright colors that enable individuals of the same species to recognize one another. The family of bee-eaters, for example, all look remarkably similar

Below The long, narrow wings of the Laysan albatross are ideally suited to using the air currents generated by ocean waves to create lift. The similarity to a glider is not accidental.

Right The peregrine falcon is the epitome of power and grace. It is the equivalent of the fighter plane and, from its prey's viewpoint, just as menacing.

and are best identified by the areas of bright color about the head. We find such color patches useful to name the various species, and there is no reason to suppose that the birds themselves do not utilize them for similar purposes.

Power-to-weight ratio
Though feathers may be the most obvious adaptation to flight, in fact the whole structure of birds is geared to providing the best power-to-weight ratio. Their bones are thin-walled and hollow and, wherever extreme strength is required, they are honeycombed. Muscle is reduced to a minimum. In fact the only really substantial areas of muscle on a bird are the flight muscles located on either side of the sternum or breast-bone. These muscles (the breast of a chicken) are divided into two sets, the upper and larger which pulls the wing downward on the propulsion stroke, and the lower which raises the wing ready for the next beat. This is the bird's "engine-room," the necessary muscle to maintain powered flight which may account for half its total weight. In birds that glide, rather than beat their way through the air, these muscles may be reduced, while in some flightless birds they are virtually absent altogether.

All of a bird's other internal organs are reduced in size according to this weight reduction axiom. The digestive system, for example, is remarkably speedy and can deal with food, extracting the useful and discarding the waste, in a matter of minutes rather than hours. The female's ovaries form eggs extraordinarily quickly. A bird simply cannot afford the luxury of carrying around its young within its body for any length of time. The weight of a clutch of eggs would prove too much of a hindrance, impede its feeding ability and make it more likely to fall prey to predators. The system of laying eggs in a nest is ideal for flying animals that need to rear large numbers of young each year to maintain their numbers.

FEATHERS

Birds are covered with feathers that evolved from the scales of their reptilian ancestors. Feathers vary according to their function. The penguins, for example, have a waterproof covering of seal-like hairs that form a "fur" to insulate them in the cold waters they inhabit. Their plumage is in marked contrast to that of the ostrich whose elaborate plumes serve a different function.

Types of feather

The feather is an acknowledged measure of lightness: "light as a feather," we say. Yet it is also the finest of insulators. The feathers of the wings that propel birds through the air are both extremely light and extremely strong. The shaft of a primary feather is hollow, flexible, but very difficult to snap. It is bordered by an interlocking series of barbs that forms an effective vane. When a bird's wing is pulled downward, the individual flight feathers overlap to form a continuous area. On the up-beat the feathers twist and open to allow air to pass freely between them.

The primary feathers (the long feathers on the outer joint of the wing) vary in number between nine and 12: they are the major propellers, the means of movement. The flight feathers on the

Below A hovering European kestrel uses its long, angular wings to maintain its stationary position relative to the ground. In fact it is moving through the air by facing into the wind. Kestrels find it virtually impossible to hover in a dead calm.

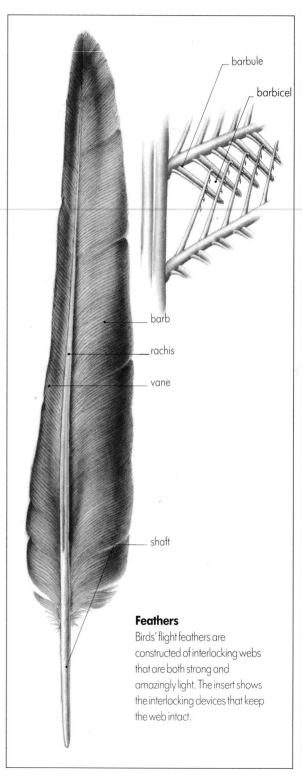

barbule

barbicel

barb

rachis

vane

shaft

Feathers

Birds' flight feathers are constructed of interlocking webs that are both strong and amazingly light. The insert shows the interlocking devices that keep the web intact.

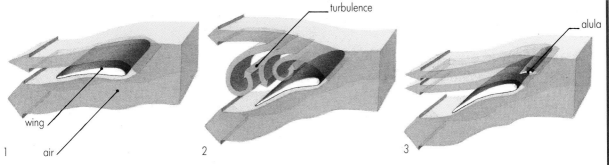

turbulence

alula

wing

air

1   2   3

Mechanics of flight

In normal flight (1) the air passes smoothly over the wing. When angled to slow down (2) the air creates turbulence over the upper wing and leads to stalling. This problem is overcome by raising the alula (3) to ensure that even at slow speeds air continues to pass smoothly.

Above Feather care is an essential part of a bird's existence. Migrants, like this whimbrel, must ensure that their flight equipment is in perfect working order before setting out on lengthy journeys.

inner joint, the secondaries, provide lift, to keep the bird airborne. The feathers of the tail help provide lift, but also act as a rudder and airbrake. The structure of both secondaries and tail feathers is similar to that of the primaries, though they are generally less stiff and more pliable, to suit their function. Both groups are also more variable in number, between different species, than the primaries. The long-winged albatrosses, for example, may have up to 32 secondary feathers.

Modern aircraft fly at extraordinary speeds, but they still have to slow down to land. The trouble is that these two requirements conflict to a considerable degree: the result has been the system of flaps that alter the geometry of the wing on a modern airliner. While fast flight requires narrow wings, slow flight requires a larger surface area. Without larger wings the plane would either have to land unacceptably fast or it would stall. Birds have their own built-in anti-stall device in the form of a bastard wing. These stiff feathers grow from the bird's vestigial thumb and have the effect of smoothing out the flow of air over the wing and reducing stalling speed.

Why birds molt

But, just as airliners have to be taken out of operation for regular servicing, so do birds have to take time off to replace worn and damaged feathers. This process is called molt. Some birds, such as the ducks and geese, actually dispense with all their flight feathers at once and may be quite flightless for days or even weeks while new ones are grown. Some of the highly colorful male ducks even change into a special camouflaged plumage, called eclipse plumage, during this highly dangerous period. Ducks and geese can stay out of reach of most predators, but most birds cannot afford either the time or the risk of such a strategy. So they molt their flight feathers gradually over several weeks, discarding the primaries sequentially from each wing starting with the innermost.

ANATOMY

Above The massive, hooked bill of the bald eagle is a tearing tool capable of dealing with thick-skinned prey far larger than this bird could ever catch for itself.

Right The narrow, serrated bill of the red-breasted merganser is perfectly adapted to gripping slippery prey under water.

Birds can be divided into a large number of categories on the basis of their food. Indeed they are adapted to take a wide variety of foods. Some are remarkably conservative in their feeding habits like the crossbill which has a bill specially adapted to opening tough pine cones and extracting the seeds. Presented with a bird, an ornithologist can tell a lot about its life simply by studying its bill and its feet. It may, for instance, have a thin insect-eating bill, or a wide gape indicating that it catches its insect food in the air. It may have the long probing bill of a wader such as a godwit or curlew, or the thick chunky bill of a grosbeak suited to cracking hard nuts. Eagles have sharply hooked bills suitable for tearing flesh, though prey is invariably taken with the sharp clasping talons. Ospreys have specially serrated talons suitable for grasping slippery fish which they hunt by swooping down to the surface of the water from the air. Another fish-eater, the red-breasted merganser, has a serrated bill for similar reasons. Woodpeckers have chisel-like bills, spoonbills along with many ducks have sifting bills, herons' bills are dagger-like and hummingbirds' fine pipe-like tubes. Each is adapted to a particular food source. But other birds are all-arounders capable of taking any opportunity that presents itself.

Starlings are particularly good examples of all-arounders, a fact which helps to explain their extraordinary world-wide success.

Feet and legs

Just as their bills vary according to food, so do birds' feet vary according to their habitats. Ducks have webbed feet and grebes semi-palmated (partly webbed) ones. Jacanas, often called lilytrotters, have long toes to spread their weight over floating aquatic vegetation, while the foot of the ostrich has been reduced to two toes suitable for running. Woodpeckers have an unusual arrangement of two toes pointing forward and two back. Such an arrangement makes a set of climbing-irons, ideally suited to clambering over consists of three toes pointing forward and one backward. It is an "all-around" foot suited to walking and grasping a simple perch, though keen bird gardeners will know that titmice sometimes grasp food in their feet like a bird of prey. Yet even among this group of "normal" footed birds there are enormous variations of structure. Pipits and wagtails have the claw of the hind toe enormously extended as an aid to balance while walking on flat ground, which they do a lot. Ptarmigan, which live at high altitudes and low temperatures, not only have a solid walking foot, but one that is completely feathered to provide insulation against the cold.

Below The powerful, heavy bill of the blue grosbeak is perfectly suited to cracking the hard shells of large seeds. In fall these birds frequently gather in large flocks in sorghum fields.

the rough bark of trees and, together with the specially stiff tail, creating a perfect tripod from which to hack forcefully during its wood-pecking activities.

Many shorebirds have very long legs enabling them to wade into deep water in search of food. Few are as long as the black-necked stilt's, which trail out behind in flight and seem remarkably inconvenient when they have to be folded up while the bird incubates its eggs. Swallows and martins have tiny legs and feet suited to grasping a thin perch but no more. Swifts have even smaller feet, with all three toes pointing forward, but armed with sharp claws that are perfect for clinging to rough surfaces.

The "normal" foot of the average perching bird

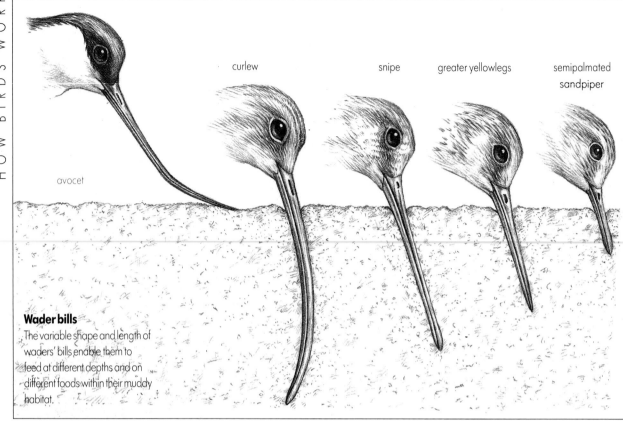

curlew   snipe   greater yellowlegs   semipalmated sandpiper

avocet

Wader bills
The variable shape and length of waders' bills enable them to feed at different depths and on different foods within their muddy habitat.

Bill–feet combinations

It is, however, when we consider combinations of bills and feet that we can find out most about particular birds. As we have seen, birds of prey catch their food in their feet, but use their bills to dismember it into edible-sized pieces. There is thus a vast difference between the bill and feet of, say, a golden eagle and one of the smaller insectivorous falcons. Similarly, although both osprey and bald eagle hunt for fish, their techniques and size of prey vary enormously. The osprey dives for fish from a height and must be able to rise from the water carrying its prey. Such fish are thus proportionately small and the osprey needs only a small sharp bill to tear off pieces to eat. Bald eagles are largely carrion eaters and, when they go fishing for themselves, they simply stand watch over a salmon-rich shallow and do little more than pounce on a fish that lands awkwardly. Such fish are often huge and it takes all the bird's strength to wrestle it from the river into the air. A tough and powerful tearing bill is required to deal with the double problem of large prey and tough-skinned carrion.

One of the most difficult bill-feet combinations to understand and interpret is that of the flamingo. This strange bird actually feeds on algae that it sieves from the shallows by pumping water through a well developed set of filters that line its bill. It also feeds with its bill held upside down, literally on its head. A flamingo's legs are extremely long, indicating that it feeds by wading in deep water. Yet because it has such long legs, it also has an extremely long neck so that it can reach the shallow mud among which it spends most of its time. Of course, flamingos do feed in deep water, but they are quite capable of swimming and even regularly up-end, duck-like, to reach the bottom to feed. It is all very peculiar, but then flamingos are peculiar birds in other ways too.

Many other birds, that have highly specialized foods or feeding techniques, are also peculiarly adapted. In fact, the more specialized a bird becomes, the more unusual its anatomy. While most North American hummingbirds have long, tube-like bills to suck nectar from living flowers, none can compare with the endowment of the sword-billed hummingbird. This bird inhabits the South American Andes and has a bill almost as long as its body. In fact, a botanist, working on the flowers of the Datura with their extremely long corollas, predicted the existence of such a hummingbird before it was actually discovered.

While many of the strangest birds tend to

The flamingo

Flamingoes feed with their head upside down and bill pointing backward. Rows of lamellae along the edge of the bill sift algae from salty water.

inhabit tropical climes, a few do venture into temperate latitudes. The snail kite of Florida, for example, has a tiny winkle-picker of a bill even smaller than that of the osprey. It is used solely for prying giant Pomacea snails from their shells. Anyone who has seen one of these snails will realize that each individual is a really substantial meal. Also confined to the southeastern United States, though found in many parts of the tropics as well, is the curiously "deformed" black skimmer. This black and white, tern-like bird is unusual in having the lower mandible considerably longer than the upper. It feeds in a quite unique way by flying low over water, skimming the surface with its extended mandible and grabbing small fish attracted to its "wake." In fact it often flies back along the same line to collect fish that have gathered during its first passage.

The bird's anatomy may be geared to flight and most are thus relatively small and light in weight, but the variety of shapes, forms, colors, and behavior patterns is more than sufficient for a lifetime of dedicated watching and study.

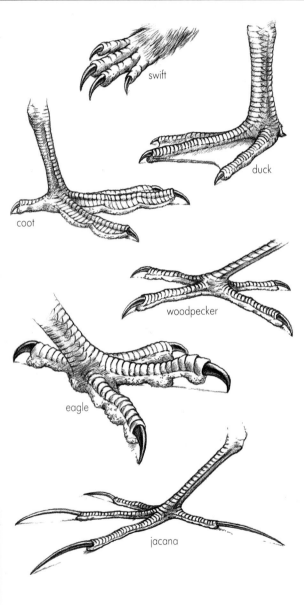

Feet

Birds' feet are adapted to a wide range of different habits and habitats. Each is suited to a particular mode of life.

THE BIRD'S YEAR

The vast majority of birds work, like ourselves, on a regular, repeated annual cycle that is considerably more complex than the "birth, copulation and death," described by T.S. Eliot as the facts of life. Some species, such as those that breed on or near the Equator, may breed all year round because the plants on which they feed bloom and fruit randomly through the year. Similarly some tropical seabirds breed at ten month intervals, the actual timing of breeding changing from year to year. But for most birds, certainly in temperate climes, life is geared to the changing seasons.

Bird watchers talk of spring and autumn migration, of the breeding season and of winter visitors, but such seasonality does not necessarily correspond with our own sense of season. Spring migrants usually do arrive in spring, but for birds in temperate North America and Europe spring can mean anything from late March till the end of May. Even so, the spring migration of European white storks takes place in January. Similarly, autumn migration can begin as early as mid-July and last through to early November. The peak period for migrating shorebirds is August, a time that most people consider the peak of the summer. And if the

Above Vermilion flycatchers are mainly summer visitors to the southwestern states, though they may frequently be seen in winter along the Mexican border.

Below Female crossbills often breed long before the snow has melted. For this bird summer may well begin as early as February – a seasonal adaptation to its food supply.

breeding season is "summer," then what is one to make of great horned owls, which routinely lay eggs in the Northeast in February? Nevertheless, there is a definite structure to the life of the average temperate-dwelling bird.

The seasonal routine

Taking the average summer visitor to our northern lands, we find that it will spend winter around the Equator and move northward in the latter part of our winter to arrive at its breeding site, perhaps in our garden, from mid-April onward. The male will arrive first and quickly establish a territory, which it will defend by displaying to its neighbors and by singing. The female will arrive a week or two later and will seek out a territory-holding male. Display will lead to mating and nest building. Eggs will be

Left In winter, gray sanderling are extremely well camouflaged among the gray mud banks of the shorelines they tend to frequent at this season.

Below left In sharp contrast they become sandy brown in summer, when they breed among Arctic wastes. This bird is in a transitional stage of molt between winter and summer.

one of the pair could die, the chicks could fall victim to a predator or the weather; but certainly by the second week of July there is a flush of young songbirds traveling through the countryside. Such movements could be regarded as the start of autumn migration.

Feeding for survival

One of the essentials for migration is fuel for the long journey to come, so the bird feeds voraciously on the plentiful supply of insects and fruit available at this time. Many small birds actually double their weight before setting off on a flight that may last 40 hours of non-stop flying over the sea, desert, or both. Finally they arrive in their winter quarters where, once again, food may be plentiful. But even here, all may be far from well.

Imagine a small bird weighing less than half an ounce flying northward on a journey of some 3,000 miles in spring; building a nest and laying its own weight in eggs; incubating these for two weeks, but still finding enough time to feed itself; then feeding a voracious brood of youngsters that, by the time they are ready to fly, are eating more than the adult itself; then feeding up to double its weight; and finally setting off to fly a return journey of the same distance again; all within a period of 12 weeks. Such a schedule inevitably takes its toll and perhaps fewer than half the birds that set off northward ever make the return trip southward. That some individuals perform this remarkable feat three, four, or even five times, hides the fact that only a small minority of the young reared in a season live to breed even once.

Understanding the bird's year is essential to all birders simply because, if we hope to see a particular summer visitor, we must see it during a few brief weeks, or wait almost a full year to try again. Similarly, birds that do no more than pass through our area have to be seen (or missed) during the brief weeks of spring or autumn. Winter visitors from lands farther north usually give us a longer period of observation potential, but only the residents are available all the year round.

laid, usually early in the morning, until the clutch is complete. They will then be incubated - for some two weeks in the case of most songbirds - and the chicks will hatch naked and helpless. The youngsters will be fed by the parents and brooded for two or three weeks more until they are able to fly and will then remain "in care" for a further few days more before drifting away. Some species will then go through the whole performance again to rear a second brood, though many will rear only one nestful.

This comparatively straightforward process takes some 40 days. That is, say, seven days to establish the pair, mate and form the eggs inside the female; a day or so to build the nest; 14 days' incubation, 16 days for the young to fledge, and three days of post-fledging care. So, for a pair of birds that have found each other by the first of May, breeding could be over by the second week of June. Of course, many things could happen to upset this schedule. The nest could be destroyed,

MIGRATION

The ability to fly makes birds the most highly mobile animals in the world. Some species like the arctic tern fly from one end of the earth to the other every year; others make more modest journeys, while yet others seldom leave their tiny territories during their entire lives.

It is a fact of geography that large areas of the world are virtually uninhabitable for lengthy periods of the year. The land beyond the Arctic Circle is a hostile place in winter that supports only a few seed eaters, other vegetarians and the predators that prey on them. In summer, however, everything changes and the tundra becomes full of life while the sun shines 24 hours each day. Many creatures such as insects survive in the Arctic to take advantage of this bloom of life by hibernating or spending the winter in a state of suspended development.

Birds, however, with one notable exception (the common poor-will has been found hibernating in rocky crevices in the Southwest) do not hibernate. Instead they migrate to these rich northern latitudes from the milder climes where they spend the winter. As a result, huge numbers of birds – insectivorous passerines and aquatic waders – perform vast migrations. The numbers involved are staggering and the losses along the way stunning. But while individuals may die, the species as a whole must benefit. It is axiomatic that if a food source exists some species of bird will have evolved the ability to take advantage of it.

The mysteries of migration

Migration has always fascinated people, but even as short a time ago as the 18th century the great migration-hibernation debate still raged. That we now know the answer does not mean that we know all about migration – far from it. We know that regular seasonal flights bring birds to our shores. We have a good idea where they winter. We know the routes some of them take and of seasonal variations in such routes. We know that most birds migrate below 4,000ft, but that others cross the Himalayas at altitudes where humans have seldom been able to penetrate without the aid of bottled oxygen. We know that many migrants are capable of covering long distances non-stop, flying from one day into the next. We know that they take on fuel in the form of fat prior to migrational flights and that some species may even double their normal weight. We know that birds of prey and storks, because they are dependent on thermals of

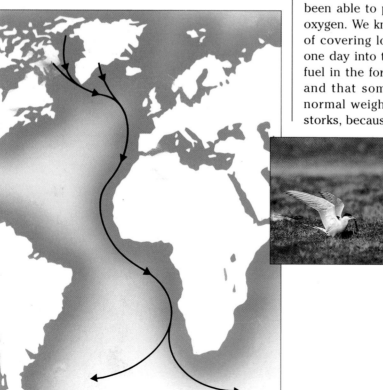

Left and inset From their breeding grounds in Greenland and the Canadian archipelago, Arctic terns make a transatlantic crossing to the coasts of Europe before heading southward to winter off Antarctica – the longest of any migrational journey.

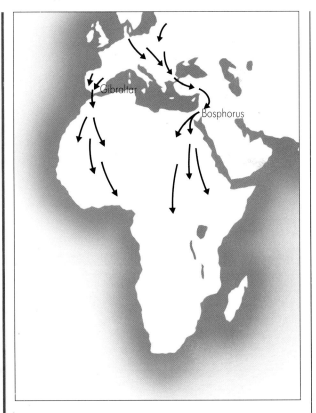

Left European white storks use thermals of rising warm air to aid their soaring migration between Europe and Africa. As thermals are not generated over the sea the birds take the shortest of sea crossings.

Below Snow geese arriving at Klamath Basin National Wildlife Refuge, California. Snow geese make use of all three major flyways during their migration through North America. In contrast, the diminutive Ross's goose crosses the Rocky Mountains between its arctic breeding grounds and its winter quarters in California.

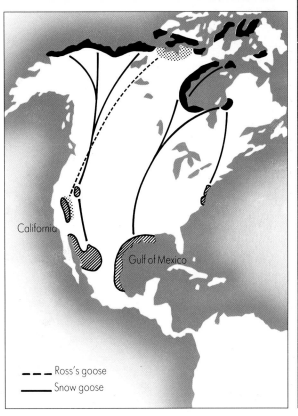

rising air generated by the heat of the sun on the land, avoid long sea crossings. And that many other birds migrate on a broad front and are not channeled along particularly favored routes. These and other aspects of bird migration we know about, but there are many questions that remain unanswered and doubtless many questions that we have not even asked.

Even birds as large and obvious as the greater flamingo pose problems. There are, for example, some 10,000 of these birds on the salt lakes of southern Cyprus every winter, but we do not know where they come from. Outside of Europe and North America the problems are immense.

Migration involves millions of birds making regular journeys of varying distances throughout the world. Some merely drop down from the mountain tops to the milder climates of the valleys or lower slopes. Such "vertical" migrations may cover only a few miles. Others, as we have seen, cover thousands of miles every year, much more than the average car, moving from one continent to another.

Inevitably, some birds get lost along the way and turn up in areas where they are completely

- - - - Ross's goose
———— Snow goose

Right Barn swallows also use wires where they are available, but in their African winter quarters these swallows are equally at home among papyrus beds.

Below Typical of their kind, tree swallows gather on telephone wires prior to migration, often in huge numbers.

unknown. In North America the island of Attu in the Bering Straits of Alaska acts as a gathering ground of birders seeking off-course Siberian birds, while in Europe the Isles of Scilly is a similar magnet for birders in search of off-course American birds. These mainly autumn rarities are but the tip of an iceberg of lost migrants. For if an individual of a species that is unknown can get lost so can individuals of a species that is regular.

Most migrant birds, however, do make their journeys more or less successfully. Leaving the area where they were raised, they fly instinctively to a traditional wintering ground, before returning once more to almost the exact area where they were hatched. Along the way they may stop off at places where they have stopped off before. So the swallow in the barn may well have hatched there, flown to a farmstead 5,000 miles away to the south, and then flown back to the very same barn.

Why birds migrate

This destination-based migration, both northward and southward, fulfills a remarkably important function. Supposing all the swallows hatched in North America flew south and then all returned on a random basis to anywhere in the United States. The result could be huge concentrations of swallows in say the Midwest, and none at all in the East, or west of the Mississippi. Such inefficiency would leave massive unexploited opportunities and staggering scarcity of food and breeding sites in the overcrowded Midwest. Even if the swallows then spread out across the country, they would be late arriving at vacant territories and be less efficient breeders as a result. By being destination bound, the swallow population spreads itself out over the suitable habitat and avoids a natural disaster.

Similar factors decree that individual swallows winter in the same area, perhaps even the same farmstead, year after year; and that along the way they roost in the same reedbed year after year. This would seem to lead us to the concept of birds having quite definite migration routes and an earlier generation of ornithologists clearly thought that birds were channeled along mountain ranges and coastlines. It does seem likely that individuals do follow the same route year after year, though maybe a different route in spring to that followed in autumn. Additionally, birds do follow mountain ranges and coastlines, but they also migrate on a broad front and take advantage of such landmarks only when they lead in the right direction.

Migration and navigation

Radar can be adjusted to pick up migrating birds and studies have shown how widespread is migration. Coasts may have a funneling effect, but migration takes place on wide fronts, taking no account of narrow sea crossings or of deserts. Small birds regularly set off from the coasts of the Mediterranean in autumn to fly nonstop across the

sea and then the Sahara Desert. Such flights must last 40 or more hours and use up an enormous amount of energy in the form of fat that they have accumulated in a pre migration feast.

Just how birds find their way is still the subject of hot debate. It is generally agreed that navigation involves the use of the sun during the day and the stars at night, for when it is completely overcast birds become disorientated and get lost. This does mean, however, that birds must have an inbuilt sense of time – an internal clock. There is also evidence that birds respond to the Earth's magnetic field and that the final stage of destination-based migration may be achieved by the sense of smell. Whatever the truth, it is clear that the whole mechanism is, for a majority of bird migrants, completely instinctive rather than learned. For while in species like the snow goose, adults lead their young on their first migration, the young of most birds find their own way to a particular spot, in a particular country, which they have never visited before.

Left During their length of the continent migrations, Snow geese fly in line and V formations, with the leading bird doing most of the work. Most migratory geese follow a similar routine.

SONGS AND CALLS

Birds vocalize for a variety of reasons, indeed they are among the most vocal of animals. They sing to proclaim their ownership of a territory, but also utter a whole range of different calls which fulfill a variety of functions. Alarm calls serve to alert other creatures, as well as their own kind, to the approach of a predator. Outside the breeding season, many species spend much of their time in groups and maintain the composition of the flock with contact calls. Some calls are uttered as threats, aimed specifically at individuals of the same species, or at intruders which may or may not be potential predators. Yet another type of call expresses the bird's excitement. There are calls that are used only during the period of courtship or mating and yet more that are confined to communication between the young and its parents. In a remarkable study, one ornithologist described no less than 57 different calls uttered by the European great tit, each presumably with a different meaning. While this should not be regarded as a language, it is a considerably richer repertoire than that of most other animal groups.

The functions of bird song

Of all of these different forms of vocalization, bird song is both the most obvious and the most pleasing to the human ear. Indeed, it is virtually the only aspect of ornithology that is regularly used in

Left Like most other woodpeckers the North American hairy woodpecker uses dead branches to drum its territorial "song." The hollow sound it produces echoes through the woods.

Above Ruffed grouse are unusual instrumentalists that use the sound of their wings beating rapidly – often while perched on a hollow log – rather than their voices to proclaim their place in the sun.

literature, showing that poets too are not immune to the charm of birds. Bird song is the means by which males (mainly) proclaim their ownership of a territory and advertise their presence to potential mates. Other animal groups perform the same functions by means of smell or by visual signals. Birds have a comparatively poorly developed sense of smell and only those species that live in open habitats have evolved elaborate visual signals. Cases in point are the ostrich, which has no song, but exhibits a wonderful semaphore-like wing flapping display that can be seen for long distances across the open African savannahs; the European great bustard, which turns itself virtually inside out on the grasslands of Spain and eastern Europe; and the prairie chickens of the American plains with their elaborate puffed up strutting.

Most bird song is relatively simple and repetitive. There is considerable evidence that many of the standard elements of a bird's song are innate, but that even simple songs may be partially learned. The complex, trilling vocalization of the male song sparrow has been shown to be instinctive; a song sparrow raised by a canary will sing the basic song sparrow melody. But for a time, between the ages of 4 and 12 weeks, a male song sparrow is able to learn variations by listening to other males, and may add some of his neighbor's flourishes to his own medley. The result is that song sparrow songs differ between regions of North America.

However, many birds exhibit a remarkable virtuosity in their songs with variable phrases and arrangements indicating an apparent choice on the part of the individual. The famed song of the nightingale is a case in point, though it is perhaps as much the quality of the notes produced by this bird as the actual arrangement of phrases that is responsible for its reputation. Several groups of birds, notably the nightjars or nighthawks, produce songs that sound more like machines than birds. They are repetitive, long-lasting, and would probably seem quite boring were it not for the secretive and nocturnal habits of these charismatic birds. Similarly many owls produce a form of hooting that is simple in form, though it is nevertheless a "song." A few birds have even become instrumentalists, using non-vocal sounds to proclaim their ownership of territory. Primary among these are the various species of woodpecker that drum on dead branches. Individual woodpeckers have their own dead branches within their territory, and indeed the presence of such "drumming" posts may be an essential ingredient in making an area suitable for breeding. In effect, the dead branch acts as a sounding box, amplifying the rapid beating with

Top The yellow-headed blackbird's song starts with a rasp and ends with a buzz. Though far from pleasant, it serves the same purpose as the more delightful songs of other birds such as the nightingale (**middle**). Curiously, some birds, such as the starling (**bottom right**), mimic other birds as part of their own song, though it is difficult to see what advantage this serves.

Above Ravens have virtually no song, but utter deep, harsh croaks that are both far carrying and distinctive. The calls of all members of the crow family are very similar and need an experienced ear to separate them.

falling tree. The northern mockingbird is also a noted mimic, though it usually confines itself to the songs and calls of neighboring birds. And even the humble starling can produce a passable version of other bird songs. It is perhaps not the individual "words," but the "language" itself that matters, in other words it is not the phrases but the accent that distinguishes one lyrebird from another one.

Birders and bird calls

So, just as one bird can recognize another (otherwise the whole business of song would be pointless), so can we, as observers, recognize the songs of birds and identify the species without seeing the individuals responsible. The practical value to the birder of knowing the songs and calls of birds cannot be overstated. A competent birder can walk through a wood full of bird song and identify most of the birds present without seeing a single individual. To the beginner such expertise seems daunting, but by starting with everyday garden birds it is surprising just how quickly one can become familiar with their songs and calls. Even by learning only a handful of the songs of the most common birds one is in a better position to pick out the more interesting species one encounters. If every call or song has to be tracked down to a confirmatory visual contact then a spring walk through a wood becomes more of an ordeal than a pleasure. Gradually, over time, expertise is built up until a wide variety of species can be identified unseen.

Probably the best method of learning bird song is in the company of a knowledgeable and understanding friend, actually out and about in the field. Failing this, recordings can be quite useful and listening to them during the long winter evenings is a pleasant and productive way of waiting for spring. Remember, however, to concentrate on the most common species rather than those you particularly want to see. Learn thoroughly, and test yourself by identifying the recordings regularly. Before long you will be moving on from woodland bird song to the calls of contact produced by a variety of species in different habitats.

Shorebirds may seem a difficult group to identify, not because they hide away in dense vegetation, but because they are often seen at considerable distance or in flight and, in any case,

the bill into a hollow rattle.

Other species produce songs by using part of their plumage. Snipe, for example, proclaim their domain by flying in a wide circle overhead and by making a bleating sound produced by vibrating their stiff outer tail feathers through the air in steep dives. Ruffed grouse in the northern woodlands produce a drumming sound by beating their wings rapidly while standing on a hollow log.

Birds use songs and calls to communicate with others of the same species. It is, therefore, difficult to see why some birds mimic others. Many species throughout the world are fine mimics producing excellent imitations of other birds and a wide variety of sounds. Outstanding is the Australian lyrebird which not only produces the songs of other birds that share its range and habitat, but sounds as variable as a tractor, a chain-saw, and a they are remarkably similar in structure. A fast-flying group of small waders may produce the rasping "kree" of dunlin, or the twittering "plick-plick" of sanderling. In poor light the call may be the only clue to their identity. Most shorebirds have distinctive calls and recognition can alert one to the presence of a species that had not previously been noticed on some overcrowded marsh or estuary.

Nocturnal calls

With some species, calls are virtually the only contact that one is likely to get without a great deal of stealth, forward planning, and local knowledge. Nocturnal species such as the owls are a case in point. Finding a long-eared owl in an extensive conifer forest, for example, is either a matter of luck, or of waiting and listening carefully for the characteristic low, moaning call. Recognizing the calls of the various species of owl is crucial to locating them, for these birds are never common and actually seeing them is always a time-consuming business. Sadly, many of the owls call only during the early part of their breeding cycle which is, in any case, usually during the early part of the year. Thus locating, say, a great gray owl involves visiting the northern forests when there is

still heavy snow on the ground and transportation is at its worst. Later in the summer these birds are quiet and secretive, making location a matter of sheer good fortune.

Similarly, nightjars and nighthawks are birds of the night, and location depends on visiting an appropriate habitat at dusk and waiting to hear their characteristic calls. Even then, one is lucky to see the bird, though several species are curious and if one waits in the middle of a suitable hunting area, a bird may well come to inspect. Fortunately, several species regularly come to roads and tracks and may then be caught in a car's headlights. It is, however, essential either to choose a seldom used road or to try very late at night when traffic is at a minimum. The exception is the common nighthawk, a familiar bird of the city, where it can be seen flying buoyantly overhead at dusk, giving its mechanical "peent" call.

Recognizing songs and calls is, then, an essential method of identifying and locating birds and a vital skill for an aspiring birder.

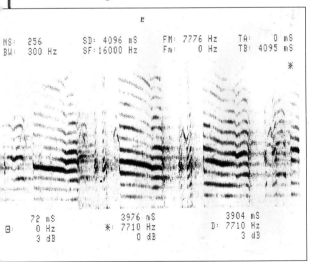

MS: 256   SD: 4096 mS   FM: 7776 Hz   TA: 0 mS
BW: 300 Hz SF:16000 Hz   Fm: 0 Hz   TB: 4095 mS

        72 mS   3976 mS   3904 mS
         0 Hz   *: 7710 Hz   D: 7710 Hz
         3 dB   0 dB   3 dB

Above Sound spectographs are a visual method of recording bird sounds and songs. This is a spectograph of the call of a kittiwake, which is rendered phonetically as *kitti-work kitti-work* and, in words, as Kittywake Kittywake.

Below Most thrushes produce rather liquid, flute-like songs that are pleasant, but fairly similar. The European song thrush has the characteristic habit of repeating each of its highly varied phrases three times. It is not what is said, but how it is said that identifies this bird.

BIRD NAMES

Identifying, that is putting a name to a bird, is the first essential in learning more about birds. However, it takes only a short time to find out that different bird books use different names for the same bird. For example, the brown creeper in North America is the tree creeper in Europe. These variations are frustrating to the beginner and irritating to the experienced.

Naming systems

Fortunately each bird not only has one (or more) English names, but also an accepted scientific name. The brown creeper and the tree creeper have the scientific name of *Certhia familiaris*.

The two-part system, or Latin binomial, involves applying a generic name (the first word) and a specific name (the second). Thus the tree creeper is a member of the genus *Certhia* and is specifically identified as *familiaris*. There are several other creepers that belong to the genus *Certhia*, but each has a distinctive second, that is specific, name that identifies it clearly and without confusion. By

convention the generic name starts with a capital letter and the specific with a small letter. Also, traditionally, the names are printed in italics.

All birds belong to the class Aves. This "class" of living thing is divided into different "orders" which, in turn, are divided into distinct "families." These families consist of different genera which, as we have already seen, consist of different species. Thus, working in the other direction, we see that the tree creeper is *familiaris* and belongs to the genus *Certhia*. *Certhia* belongs to the family Certhiidae which forms part of the order Passeriformes which is part of the class Aves.

One of the great things about the system of scientific names is that it enables birders from different countries to converse freely and easily

There are only five species of flamingo in the world, yet they demonstrate just how a group of closely related birds are named in the scientific system.

Left Greater flamingo

Above Chilean flamingo – both are *Phoenicopterus*.

one with another. Latin names overcome the difficulty and frustration of spending ages thumbing backward and forward through field guides to point at illustrations of birds we wish to discuss with foreigners.

With luck, you have now got the hang of Latin binomials: now comes another minor complexity – Latin trinomials. The basis of our classification system for birds is the species, but there are variations in color, voice, and behavior that enable

Orders and families

The flamingos of the world may, however, make the business of names a little clearer. Firstly, all flamingos are placed in the order Phoenicopteriformes. Secondly, they are all members of the family Phoenicopteridae. They are thirdly divided among three distinct genera: *Phoenicopterus, Phoeniconaias* and *Phoenicoparrus*. There are, in fact, only five species of flamingos: Greater flamingo *Phoenicopterus ruber*, Chilean flamingo *P. chilensis* (note that the generic name

Far left Lesser flamingo at its breeding mound. This bird is placed in a genus by itself – *Phoeniconaias*.

Left Andean flamingo.

Below Andean and James's flamingoes feeding side by side. These two birds are so closely related that they are placed in the same genus *Phoenicoparrus*.

us to pick out subspecies. In most cases, this has required detailed comparison of a large range of preserved bird skins in a museum, but some subspecies are sufficiently distinct to be recognizable in the field.

The various populations of the greater flamingo are sufficiently different to merit status as subspecies. Indeed, not so long ago, two were regarded as quite distinct species: there was the greater flamingo *Phoenicopterus roseus* of the Old World, and the American Flamingo *Phoenicopterus ruber* from the United States, the West Indies, and the Galapagos Islands. Today, both populations are regarded as subspecies of a single species, the greater flamingo *Phoenicopterus ruber* : the Old World greater flamingo being *Phoenicopterus ruber ruber,* and the American flamingo *Phoenicopterus ruber roseus*. If this is all a bit confusing, I make no apologies. Senior ornithologists are continually debating the relationships between different species, sometimes suggesting that they should be placed in one genus, sometimes in another. Like subspecies, these arguments should probably be ignored until the individual finds them interesting.

can be abbreviated to the intial letter and a period on a consecutive mention), lesser flamingo *Phoeniconaias minor* (note that having only one member, this genus is referred to as being "monotypic," that is a one-member genus), Andean flamingo *Phoenicoparrus andinus,* and James's flamingo *P. jamesii*.

BIRD BEHAVIOR: TERRITORY

Territory is the mechanism by which birds space themselves out over the available habitat, and the possession, or ownership, of a territory is essential if an individual bird is to breed. In our gardens, parks, and countryside, spring sees individual birds, usually males, dividing up the land among themselves. They sing from prominent positions to inform other males of their ownership and display to, or even attack, intruders. Individuals that do not gain a territory are doomed, not only to not breeding, but to a life of continuous harassment, or an existence in marginal habitats not really suited to their needs.

actually need a territory large enough to provide all their food. So we can think of the territory as a breeding space, an area where the pair can build their nest, lay their eggs, and rear their young in peace. It could be argued that the number of birds is dependent on the number of suitable territories available, though other ornithologists argue that it is the amount of food that determines the population of a particular species. Whatever the answer, it is clear that territory is crucially important to the success of a new generation.

Territories vary enormously from one species to another. The golden eagle needs a range of many

Territorial essentials

For small birds the essentials of a territory are a choice of nesting sites and a suitable song post or two. Strange as it may seem, relatively few birds

Below The territories of a robin have been mapped in this riverside suburb. Naturally the birds do not dispute the river itself, although one lives on an island and shares a small corner with a neighbor.

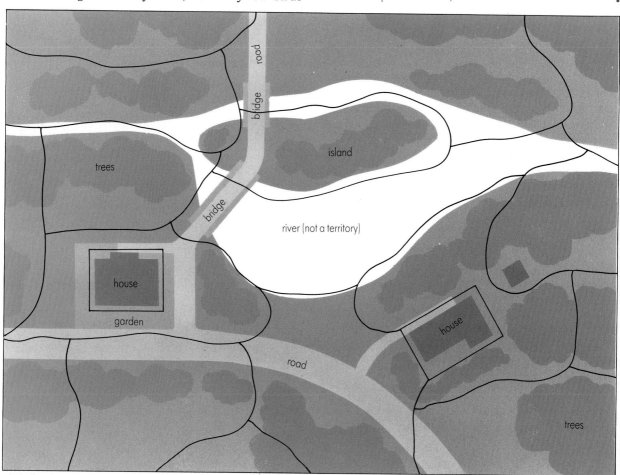

Left Male blackbirds are among the most territorial of birds and will even divide a garden lawn between them. Just what distinguishes one territory from another in an area so apparently devoid of landmarks requires a glimpse into the blackbird mind.

Below Two great crested grebes come together in a ritualized display that forms an essential part of the breeding routine of these totally aquatic birds.

Bottom Gannets are highly aggressive during the breeding season.

square miles, with three or four quite separate nest sites, to fulfill its needs. This large and powerful predator defends not only the area around its nest, but also a hunting range that is itself large enough to support a self-sustaining population of grouse, hares, and other prey. In contrast, the gannet defends a territory only as large as it can reach with its bill while sitting on its nest, that is little more than a square yard. These birds too are large and powerful predators, but they find their food away from the nest site, at sea.

There are, however, many other examples of territories that fit no pattern at all. Many of the grouse have communal display grounds called leks where males gather, often morning and evening, to attract a mate. Each lek is divided into mini-territories owned and defended by an individual male. The possession of a territory at a lek is essential to any breeding male and the more mature and experienced males will gain the better territories right at the center of the display ground. In this case, it is the female that chooses the male and it may happen that the dominant male, at the center of the lek, is chosen over and over again by different females. Having mated undisturbed by rival males, the females depart to rear the young by themselves.

Competing for territory

Outside the breeding season many birds join in flocks and coexist quite happily, but some maintain a territory throughout the year. The European blackbird is a good example, for while many northern populations migrate and roam their winter quarters together, those that are resident in milder climes are aggressive throughout the year. Competition among these blackbirds is fierce and boundary disputes occur all year round. If one individual dies, its place is quickly taken by another which has to learn exactly where the boundaries have been drawn. The only way to do this is by trial and error, so the newcomer pushes outward from the center of its newly found territory until its neighbors weigh in with vigorous

Left Birds of open grassland rely on elaborate displays rather than on song to attract a mate. The great bustard is capable of turning itself virtually inside out to create a visual effect that can be seen at great distances.

Above Sage grouse too live in open areas, but males gather at special jousting grounds, or leks, where they display to great effect with tail raised and breast puffed out, to attract a mate.

bouts of song and display.

The value of such traditional territorial boundaries may be important, not only to the individual, but also to whole populations of birds. In 1962-63 the British Isles experienced one of the coldest winters in recorded history and millions of birds died. However, with the coming of spring everyone expected that the survivors would enjoy a bumper breeding season by having the usual amount of food to share between fewer individuals. Yet, strangely, the smaller population actually had a worse than normal breeding season. Instead of mating, nest building, and getting on with rearing young, blackbirds spent almost the whole of the summer fighting and displaying. What had happened is that so many individuals had perished that the communal memory of traditional territory

boundaries had been lost. Instead of a recovery, the blackbird population barely managed to sustain itself until the next year.

Territory is, then, a feature of the lives of many birds and is essential to their successful breeding. It is, also the means by which birds will spread themselves out through the countryside and by which every suitable area is occupied.

Plumage and display

Grouse gathered at a lek do, as we have seen, perform elaborate displays both to defend their territories and to catch the eye of a potential mate. Such displays make much use of particular groups of feathers such as the tail, which is spread or fanned in a large number of species. The European blackcock and the capercaillie make much of their tails as do the ruffed and sage grouse of North America. The turkey too has a fine tail display, but none can compare with the amazing tail display of the peacock of southern Asia.

Yet even quite small and dully colored birds do all they can to show their finery off to best advantage. The ringed plover of Europe and semipalmated plover of North America are very similarly plumaged, sharing a fine disruptive camouflage that makes them difficult to see on their shingly breeding grounds. Yet their displays make much use of their banded tails which are spread and raised at any intruding plover.

Some of the most beautiful, elaborate, and bizarre displays are those between a pair of grebes. It was the British scientist Sir Julian Huxley's pioneer work on great crested grebes in the early 20th century that was largely responsible for the birth of the new science of ethology (the study of animal behavior in the wild). These attractive, lake-dwelling birds perform an elaborate series of "dances" that are designed to bring both members of the pair to the point of readiness to mate and to bond them together for the period of rearing the young. So elaborate are some of these "dances" that Huxley gave them individual names and yet they pale in comparison with the "dances" of the American western grebe. A pair of this species will come together and then, rising from the water, actually run side by side, tango-style, for many yards over the water's surface.

Left The European robin is among the most aggressive of all birds in defence of its territory. This bird is attacking a stuffed robin that has been set up to provoke displays that make much of the red breast.

Below The capercaillie is a huge turkey-like grouse of the dense conifer forests of Europe. Males have elaborate displays, but are also highly vocal in their efforts to attract a mate.

NESTS

It is easy to think of a bird's nest as a simple, neat, cup-shaped structure made of grasses, lined with a little wool or a few feathers and placed among the twigs of a tree or shrub. In fact, many birds do construct such a nest in which to lay their eggs and rear their young. But just as birds vary in size, coloration, food, and habitat, so do their nests vary enormously. Some are intricately woven, miniature masterpieces of construction with secret entrances and the capacity literally to expand to accommodate a growing family. Others are mere depressions in the ground, totally devoid of lining or decoration. The basic requirements of any nest, be it complex or simple, is that it holds the eggs and young safely.

Nesting tactics

With safety as the key, we can more readily understand the various tactics that individual species of birds have evolved in the construction of their nests. The nest is only one factor in the breeding equation. Others are the color of the adult, particularly the adult that performs the incubation; the color of the eggs, and the color of the young. If we take a bird that lays white or lightly colored eggs, then the other variables must, in some way, compensate for their extreme conspicuousness.

The bird that lays white eggs has various options open to it to protect its embryonic young. It could lay them out of sight in a hole, it could construct a domed nest, or it could lay them quite openly and cover them with its own highly camouflaged plumage. In fact, various birds have taken up all three options. Woodpeckers excavate holes in trees, kingfishers holes in banks, and storm-petrels holes in the ground or rock crevices. The winter wren has adopted the second tactic by constructing a domed nest, while nightjars and nighthawks rely on their own amazing camouflage to cover their speckled eggs laid quite openly on the ground.

Taking the other extreme of birds that lay highly camouflaged eggs, there is hardly any type of nest construction that is unsuitable though laying camouflaged eggs in a dark hole not only has no advantage but, because they are difficult to see, may actually be a disadvantage. Clearly, camouflaged eggs are best suited to birds that nest openly on the ground, or to birds that build open nests in vegetation and which, being brightly-colored themselves, must abandon their eggs to

Above A pair of song thrushes at their sturdily built nest constructed of grass stems reinforced and lined with mud. These birds will frequently build one nest on top of another, but never reuse an old nest.

Right A robin feeds its six-day-old youngsters in a similar nest, but one which is lined with grass stems.

least tern. In fact, while terns lay camouflaged eggs mainly on the bare ground, they are very gregarious and rely on aggressive communal displays to drive predators away.

With cup- or platform-nesters the first ingredient is to ensure that the nest itself is well hidden and camouflaged. The delicate, lichen-covered nests of some of the finches are highly effective in avoiding the eye of predators. Birds such as these are often quite boldly colored, but the female can be dull in comparison and, therefore, a more effective nest cover. It is not surprising that where one member of a pair is more dully colored, it is that bird that performs, if not all then at least most of, the incubation duty.

Non-nest builders

Ground-nesters apart, there is one other group

Above The nest of the colorful northern oriole is suspended in the outer twigs of a large-leaved tree. It is a tightly woven, purse-shaped structure with an entrance at the top.

Right The ringed plover lays its eggs on bare ground where they are protected by their highly cryptic coloration.

protect themselves from danger.

The eggs of ground-nesters are often so well camouflaged that the addition of nesting material would actually reduce their effectiveness. It is quite dangerous to search for such nests because of the danger of stepping accidentally on the eggs and many nests are destroyed in this way every year. Yet even with such an effective form of concealment there are minor tactical variations. The camouflage of the small plovers is highly disruptive and effective, yet the nest and eggs bear a marked similarity to those of the boldly white

that constructs no nest at all. These birds simply take over the disused nests of other species and, with little or no repair work, lay their eggs in another's home. Several birds of prey fall into this category, but one species is quite remarkable. The red-footed falcon is a gregarious summer visitor to eastern Europe that utilizes nests vacated by rooks. In general rooks breed early and the falcons arrive late. Sometimes, however, perhaps because of a particularly hard winter, the rooks postpone breeding for a week or two and the falcons arrive to find young rooks still present in the nests.

Inevitably battles ensue, for the rooks are reluctant to leave and the falcons are aggressive in their attempts to persuade them to do so.

Cuckoos, (see page 49) have their own peculiar no-nest solution to the problem.

Just as the tactics of nest site selection and nest placement vary widely, so too do the methods of construction. Some birds nest on the ground in splendid isolation, while others do so gregariously, with nests packed tightly together. Some excavate a hole away from it all, while others prefer to honeycomb a bank with nest holes cheek by jowl. Some hide their dainty little nesting cups in the base of a bush or beneath a clump of grass, while others build packed together for all to see among the highest branches of the tallest trees. Some birds weave their nests from fine grasses and mosses held together by spiders' webs. Others plaster the structure together, grasses and mud forming a primitive reinforced "concrete." Some use the same nest site year after year, gradually forming a veritable mountain of vegetation, while others seem to be content with a handful of sticks.

Frankly, I've always found the nesting habits of birds that make do with the skimpiest of nests something of an enigma. The pigeons and doves are prime examples, for they seem almost unconcerned for the safety of their eggs. A few twigs wedged together to form a platform and, bingo, they lay their eggs. Breaking all the rules these are boldly white and obvious to any predator. While the birds themselves are not boldly colored, neither are they effectively camouflaged. Perhaps the pigeons were once all hole-nesters as several species such as the rock dove still are today. But pigeons have increased dramatically by virtue of our changing the landscape for food production and, perhaps, these increasing numbers just could not find sufficient holes in which to nest. This is all very speculative, but it is difficult to explain their strange nesting habits in any other way.

Right Like many other birds of prey, the broad-winged hawk regularly adds fresh greenery to its nest throughout incubation and rearing its young.

hole-nesters like the bank swallow, and mud-builders like the barn swallow. Their eggs are white with, in most species, some sparse speckling. While the barn swallow constructs a neat cup against the wall of a barn or outbuilding, bringing individual pellets of mud to be reinforced with grasses, several other species build either complete globes or flask-shaped constructions beneath horizontal surfaces. In all cases the art of construction involves waiting for one course of mud "bricks" to dry before adding the next. This, in turn, means that the birds must have a reliable source of mud fairly close to hand. In areas with summer rain, the birds may rely on mud formed by puddles, but in dry areas a pond, stream, or river nearby is essential.

Finding out about the nests can be fascinating, but it can also be dangerous. When inspecting a nest, obey the law and ensure that your enjoyment has not left telltale signs that would draw the attention of predators.

Above The nest of the long-tailed tit is a masterpiece of grasses and lichens bound together with spiders' webs, which expands as the large brood of chicks grows in size.

Right The nest of the calliope hummingbird is a delicate little cup, covered with lichens that effectively hide its contents amid the damp forests and tundra that it inhabits.

Another strange, though more easily explained, nesting tactic is that adopted by many of the grebes. These are among the most aquatic of all birds, finding everything they need in the waters on which they live. The nest consists of a floating pile of vegetation anchored to some emergent vegetation or overhanging shrub. Like an iceberg, the structure is considerably larger below the water than above and is essentially a floating compost heap. As the water level varies, so the nest rises and falls with the flood, an important safety feature when the fragile eggs are no more than an inch or two above the water line. But, being composed of vegetation, the pile quickly starts to decay and generate heat just like a garden compost, so the eggs are warmed in a natural incubator. Like other birds, grebes incubate their eggs, but when they leave the nest, the eggs are covered with vegetation from the rim of the nests. This hides the eggs from predators, as well as staining them a mottled brown.

The swallows and martins are divided between

EGGS

The egg is the means by which birds produce embryonic young without being so weighed down that they are unable to fly efficiently. Many small birds lay their own weight in eggs to produce a full clutch. Were females to carry around such a load prior to giving birth to living young, their powers of flight and feeding ability would be seriously impaired. So the egg is another adaptation to flight.

Left During the pesticide fiasco of the 1960s, peregrines were seen to eat their own eggs – the shells were so thin that the birds were feeding on the contents of broken shells.

Above Like all other plovers, killdeer lay well camouflaged eggs that are well suited to the dry grassy plains these birds prefer.

Smallest and largest

Eggs vary enormously in size, number, and coloration. The largest egg in the world is produced by the ostrich and weighs in at about 3lb. It is the world's largest single cell, yet it is dwarfed by the egg of the now extinct elephantbird of Madagascar, which produced an egg of some 25lb – about the same as 200 chicken eggs. Yet this massive egg weighed only two to three percent of the weight of the bird that produced it. In contrast the New Zealand kiwi lays a single egg that is almost a third of its own weight. Hummingbirds lay the world's smallest eggs, with the Cuban bee hummer producing an egg of only a few hundredths of a gram.

Many birds lay one egg each day until their clutch is complete, but most birds cannot count. Thus if eggs are regularly removed from the nest the female will be induced to lay more than normal in her efforts to make up a full clutch. As a result, gamebirds such as the pheasant can be robbed by gamekeepers and the eggs placed under a broody hen or in an artificial incubator, so increasing the number of young that can be reared for sport. A single hen pheasant may, by this method, be persuaded to produce up to 30 eggs in a season. Yet the tiny golden-crowned kinglet produces a clutch of 10 eggs, weighing one-and-a-half times its own body weight, without any artificial encouragement at all.

While some birds such as the pheasant and kinglet produce large clutches of eggs, others produce much smaller clutches but are capable of rearing more than one brood in a season. Residents generally start breeding earlier and finish rearing later than summer visitors. Some of the more common thrushes, for example, produce three, four, or even five clutches of eggs each year. At four or five eggs per clutch, a total of 20 or more eggs may be laid in a single season. Other species, such as the fulmar, produce only one egg each year, yet they still manage not only to maintain their numbers but enjoy a population boom.

Just as birds' eggs vary in size and number, so too do they vary in color and shape. In some cases it is easy to see why the eggs of a particular species are shaped as they are. A glance at the eggs of a ground-nesting shorebird, such as the

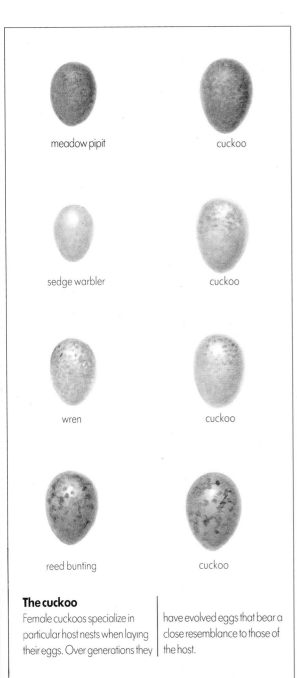

meadow pipit

cuckoo

sedge warbler

cuckoo

wren

cuckoo

reed bunting

cuckoo

The cuckoo

Female cuckoos specialize in particular host nests when laying their eggs. Over generations they have evolved eggs that bear a close resemblance to those of the host.

hundreds of feet up a sheer cliff. The color of eggs, varies from white and boldly colored to blotched, scrawled, and heavily camouflaged.

Uninvited guests

Among the world's birds, there are about 80 species that lay their eggs in the nests of other birds and leave incubation and rearing to foster parents. Yet though we tend to link such brood parasitism with the cuckoos, only just over a third of the world's 125 cuckoos actually behave in this way. In fact, none of the North American cuckoos is brood parasitic, though there are other birds that have adopted the parasitic strategy. Foremost among these are the cowbirds. The brown-headed cowbird has been recorded laying its eggs in the nests of over 200 different species though, like the Eurasian cuckoo, it does have distinct preferences. The cuckoo, has laid its eggs in the nests of no less than 300 species, though it too specializes in a relatively few common species. In fact, female cuckoos tend to specialize in nests of a particular species and lay up to 15 eggs that bear a strong resemblance to those of the host.

Below Many ground nesting birds, like the European woodcock, lay camouflaged eggs. When they hatch, the white interior of the shells spoils the effect, producing a moment of danger before the young move away.

black-bellied plover, shows that the four eggs fit neatly together with points toward the center like the segments of an orange. With birds that lay many more eggs than that number, the advantage of such a shape quickly disappears. The common murre lays a single egg that is so sharply pointed that it spins rather than rolls, a great advantage to a bird that lays its egg on a tiny ledge often

INCUBATION AND REARING THE YOUNG

Above Many ground dwelling shorebirds, like this killdeer, will perform elaborate displays of injury feigning to draw would-be predators away from their nest and eggs.

Right The burden of forming and incubating the clutch of eggs places a heavy strain on many birds. This male robin feeds his mate partly to nourish her, but also to reinforce their bond.

The objective of a bird's nesting routine is to raise as many young as possible to independence. The choice of nest site and nest construction; the timing of breeding; the number, color, and shape of the eggs, and even the incubation routine are all geared to this end.

Incubation routines

As with other aspects of birds' lives, incubation routines vary enormously from one species to another. Most species develop an incubation patch during the breeding season and, while mostly it is the females that have this patch of bare skin on their breasts and bellies, in some birds it is only the male, and in others both sexes show this feature. The presence of a brood patch generally indicates the role of the sexes in incubation. In some species males may lack a brood patch and take no part in incubation at all, though they may or may not contribute to feeding and caring for the young. In other species, they may take an equal share while in some, like the phalaropes the male may take sole charge of the chores of incubation and rearing the young. Such role-reversal, as it is

called, is a well developed tactic among many of the shorebirds that breed in the high Arctic where the season is so remarkably short. With the phalaropes it is the female that is more boldly colored, that takes the most active role in courtship and leaves her eggs in sole charge of the male. Meanwhile she is free to find another mate and produce another clutch of eggs, thus maximizing the chances of rearing young.

Some species, such as the owls, start incubating as soon as the first egg is laid, but many others do not start until the clutch is complete. The latter routine, adopted by most ground-nesting birds as well as by many small birds, has the advantage that the young all hatch at more or less the same time. With ground-nesting birds such as the

shorebirds and grouse, the young are active soon after hatching and can leave the nest in search of food and to avoid danger. If the young hatched over a period of a week or more, the adults would have the problem of guarding the active young while continuing with the incubation of the remaining eggs; clearly an impossible task. Most small birds hatch naked, blind, and helpless and must be fed by their parents for two or three weeks in the safety of the nest. Synchronized hatching, that is all the eggs hatching within a few hours of one another, produces a brood of similarly sized young. With owls, the eggs hatch over the same period of time that they were laid, producing a brood of different-sized youngsters. The eldest gets the lion's share of the food and is completely satisfied before the next largest and strongest is fed. This tactic means that if food is abundant the owls will rear many young, but if it is short only the strongest will be fed and survive. The later hatched owlets will simply die of starvation. Though it is less obvious, small birds have a similar tactic. The adults feed the largest and most available mouth each time they visit the nest, rather than sharing food equally.

This seems remarkably cruel and unfair yet it has the advantage that, when food is in short supply, at least one or two young will be reared, rather than the whole brood dying for lack of food.

The strain of incubation

A great many birds share the chore of incubation so that each gets time off away from the nest to feed, drink, and bathe. Nevertheless, and despite the long days of summer, the process places an enormous strain on the adults. During the early part of incubation the sitting bird may leave the nest quickly and discreetly at the slightest sign of danger. The stimulus of danger is much more powerful than the drive to incubate. But, as the eggs get closer and closer to hatching, the drive to incubate grows in strength and the sitting bird

becomes progressively more reluctant to leave the nest. At this time birds may appear remarkably tame, allowing a closer approach than at any other time of the year. Some Scandinavian waders have returned to incubate their eggs held in a human hand, even when the hand is lifted from the ground. An exceptional case occurred in Scotland when a red grouse was found burnt to death on her clutch of eggs after a moorland fire.

With most small birds incubation lasts about two weeks, but the young Eurasian cuckoo hatches in 11 or 12 days so that it is older than its siblings and can eject them, before or after hatching, and have the nest and its foster parents to itself – essential for so large and voracious a youngster. Other species have longer incubation periods, with most eagles having to incubate their eggs for six or seven weeks, while the larger albatrosses have to incubate for up to 10 or 11 weeks.

Left After three weeks or more of incubation this young oystercatcher has started to break out of its eggshell. The white spot on its bill is a special egg-tooth that is lost soon after hatching.

Above Laid on its back, a marsh tit will show the bare skin of its brood patch when a few loose feathers are lifted.

After a variable but always lengthy, period of incubation the eggs hatch to produce young birds. They may vary from the naked, blind, and helpless chicks of many typical garden birds to the downy and camouflaged young of the shorebirds and grouse. In both cases the behavior of the parents undergoes a radical and rapid change. The stimulus system of eggs-incubate, becomes young-brood, then begging mouth-feed.

Many young birds have remarkably colorful gapes (mouths) often spotted with contrasting colors. The response of the adult is to thrust food at the spots. It is a simple stimulus-response situation. Generally, the larger the mouth the more food is thrust into it which, as we have seen, has the effect of satisfying the strongest and most advanced chick at the expense of its siblings. The work rate of adults feeding a brood of hungry youngsters is often prodigious. Literally hundreds

Above Young woodpigeons are unable to digest the grain and seeds that form the diet of adults, so they are fed on "pigeon's milk" secreted from the lining of the parent's stomach.

Left Garden warblers are summer visitors to Europe that time their arrival and breeding activities to coincide with the hatch of green caterpillars of the various winter moths.

of sorties in search of food are flown every day in the effort to satisfy an ever hungry brood.

Hatching times

The timing of hatching is often critical in many species, particularly those that rear only a single brood each year. The European blue tit is a typical example. It lays some seven to 12 eggs over a period of, say, 12 days, incubates for 12 to 16 days, and then feeds its young on the green caterpillars of the winter moth. Should the birds get their timing wrong, and the flush of caterpillars not be

available, there is little or no chance of their being able to rear a successful brood. Yet, despite the fact that the caterpillar hatch can vary in the time of its appearance from year to year, the blue tits get it right season after season. Just how they manage this is still a matter of conjecture.

The tactic of feeding young at the nest is adopted by a huge number of different species and, in many cases, involves the adults in a remarkable change of diet. Many seed-eating finches change to insect hunters, either during the entire summer, or at least to feed their young.

Species that cannot adapt to such a change are unusual, but do include the pigeons and doves which feed their young on a liquid substance called pigeons' milk that is secreted from the lining of their crop. Clearly, this is far more nutritious and more easily digestible than a diet of hard grain seeds.

The young on the nest

Most birds carry food to their young in their bills and, during this phase of the breeding cycle, nests are particularly easy to locate. It is, therefore, a time for caution by the would-be nest watcher. A few species actually eat the food collected and return to the nest with empty bills. Young cormorants are particularly adept at getting their parents to regurgitate partly digested food for them to eat and, late on in the rearing process, will actually thrust their bills and heads deep inside the adult's gullet to produce the goods. The pelicans have a similar process, though in their case the food is regurgitated into the adult's bill pouch which acts as a sort of plate.

Eagles and other birds of prey carry food in their talons and arrive back at the nest with items that are quite beyond the chicks' ability to deal with. The adult will then tear tiny morsels from its catch to feed delicately to its young. As they grow in size, power, and appetite, such tendernesses are abandoned and prey is simply dumped at the nest for the youngsters to feed themselves. Some quickly learn to tear up their own prey, but there is often an intermediate period during which quite large items may be swallowed whole. Young owls have been timed taking up to an hour or more as they struggle to down a substantial rodent.

Many ground nesting birds produce active young that quickly leave the nest and feed themselves. Mallards on the park pond are a good example, but ducks, geese, and most shorebirds follow a similar routine. Some of these species, particularly the various species of diving ducks, are hole-nesters and, utilizing old woodpecker holes, the young may find themselves high above the ground. This is, however, no deterrence and at only a few hours of age the whole brood will throw themselves one after the other into the air to reach their encouraging parent below. They do, of course, bounce, but accidents are surprisingly rare.

Adults of species that produce self-feeding young still have to brood and warm their offspring as well as protect them from predators. Most of these species have a distinctive alarm call that elicits an instant and immobile crouching on the part of the chicks. They may scatter and crouch and rely on their camouflage to avoid becoming food to another species. Some, such as the plovers, have evolved an elaborate injury-feigning display by the adults to draw the attention of the predator away from their young. Such distraction displays often involve what appears to be a broken wing dragged over the ground.

Below Mistle thrushes breed early in the year, often before the trees have come into leaf. For this reason the nest is usually placed in the major low fork of a large tree.

LIFESPANS

It is a reasonable assumption that the larger the potential number of young an individual species is capable of rearing, the shorter the lifespan the fledgling can expect. In other words, species that can replace and maintain their numbers by rearing only a single youngster a year must live longer than those that rear 10 or 12. It is also reasonably obvious that if we notice about the same number of individuals of a particular species year after year, but know that the species concerned lays, say, two clutches of five eggs each, then an awful lot of eggs, chicks, and fledglings cannot have survived their first year.

The survival of the fittest

In fact, most small birds do not survive to breed at all; but the longer they survive the longer still they are likely to live. The most vulnerable part of a bird's life is in the nest, for whole clutches of eggs are lost to predators, bad weather, and accidents; and as a nestling the youngster faces the same

Left Ducks and other waterfowl produce large broods of young, but rear relatively few to fledging. Once they can fly these birds have a reasonably good chance of surviving to breed the following year.

Above The average American robin can expect to live a little over a year, once it has survived the first few dangerous weeks of life. In general, more mature birds survive better than inexperienced ones.

threats and is equally vulnerable. Eventually, it fledges, leaves the nest on poorly developed wings and is quite incapable of feeding and fending for itself. The death of a parent at this and any earlier stage can be a disaster.

Having become independent, the young bird has to learn the ways of the world, perfect its skills and build on the inherited knowledge with which it was born. Many will die before they acquire such skills and, in the case of migratory species, many more during their first long journey. Resident species will have to find sufficient food to see them through the winter, for it is the leanest time of the year that causes the death of most adult birds. With small birds, those that survive will breed in their first year, but with larger birds adolescence may last two, three, or even seven or eight years.

This period of immaturity is often spent well away from the breeding area, usually in the wintering area. Ospreys, for example, spend their first winter in the area where they will winter as adults. They may stay on for the subsequent summer and a second winter, but many will move northward during their first spring and over-summer in between the breeding and wintering grounds. During their second summer they may move even further northward and actually visit, but not breed in, their eventual summer homes. Such a gentle upbringing through adolesence may be an essential ingredient in their eventual success as breeding adults.

The larger albatrosses of the southern hemisphere enjoy a considerably longer period of immaturity. A youthful royal albatross cannot expect to breed until it is eight years old and may not even do so until its eleventh year. This extended period is probably necessary to enable the bird to become expert at finding its food and dealing with the vagaries of the southern storm belt. The young albatross will circle the world over and over again during adolescence. Even when fully mature the great albatrosses breed only in alternate years, so long is the period of nesting, incubation, and care of the young. Clearly, such birds can expect to reach a ripe old age. We certainly have records of one royal albatross that was banded when at least eight or nine years old in 1937 and was still breeding in New Zealand in 1988 at over 60 years of age.

Life expectations

For most birds, however, life is short. The average mature individual – that is a bird that has survived all the dangers of being reared and successfully fledged – may still enjoy a life expectancy of a year or less. The mature American robin can expect to live a further 11 months, a starling 18 months, a woodcock over two years, and a swift over three years. Most deaths are the result of accident or predation, for very few birds indeed ever die of old age in the wild.

The influence of humans on a bird's life may be particularly important, and most game birds live relatively short lives. The average adult mallard cannot expect to live another year and most grouse share that fate. Generally, these birds produce large clutches of eggs and have the ability to make speedy recoveries from population crashes. The ban on shooting brent geese in some areas of Europe has resulted in an enormous increase in their numbers, showing that the previous decline was largely due to conditions on their wintering grounds.

Species that reproduce only slowly, such as the albatrosses and most larger birds of prey, cannot compensate for persecution and may take many years to recover their numbers. Some, like the California condor, may reach such low figures that they will never be able to recover.

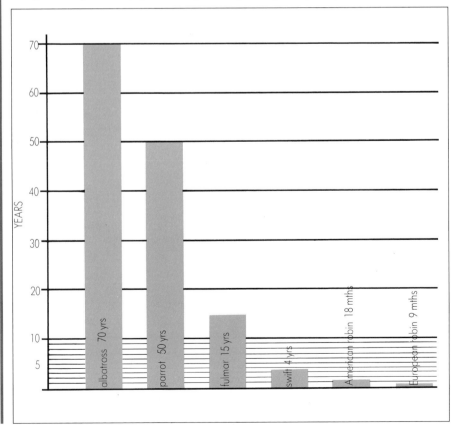

Left The average lifespan of various species is correlated with the age at which they breed; the number of eggs laid; and the number of young reared. Many small birds do not survive their first year.

YEARS

albatross 70 yrs
parrot 50 yrs
fulmar 15 yrs
swift 4 yrs
American robin 18 mths
European robin 9 mths

LOOKING OUTWARD: BINOCULARS

Knowing the birds that live alongside us, that share our home, or at least our yard, and understanding something of the ways in which birds live, we are now ready to start looking outward to the birds that live beyond the back yard. Some species will never alight in our gardens, even if they sometimes fly over, and it is part of the natural progression to becoming an ornithologist or birder to want to see new birds. At first the local park may produce a few new birds, particularly if it offers habitats, such as a pond, that do not exist in the yard. Then we may hear of a nearby woodland, marsh, estuary or, if we are lucky, a nearby bird reserve. Visits at any time of the year will start producing new birds, birds that we have previously seen only illustrated in books or magazines. But just as we expect to see different birds in the yard at different seasons, so can we expect different birds to appear in these other habitats on a seasonal basis. Some people become so involved in their garden that they seldom venture much further and satisfy themselves by producing even better bird

Above A group of birders on the Isles of Scilly, England, with every pair of binoculars focused on some poor transatlantic waif. The total value of the optical gear brought to bear amounts to tens of thousands of dollars.

conditions in their own back yard. Similarly others find a particularly satisfying habitat or area near their homes and become dedicated to their "local patch." Most of us enjoy both of these types of birding, yet still wish to travel progressively further afield. Whatever our preference, we all need equipment to help us enjoy our hobby to the full.

Binoculars – a birding essential

It was often said (and written) in the past that watching birds was one of the cheapest of all hobbies. It required no equipment, just an inquiring mind. This is (and was) untrue. Watching birds requires at the very least a pair of binoculars – or field glasses as they were previously called. To try to enjoy birding without optical aid is to doom oneself to endless frustration.

There are, however, binoculars and binoculars; the beginner should beware and follow the advice of experienced birders. Birders have several major requirements in a binocular that are different from those of other users. Coastguards need powerful magnifications so that they can identify individual ships at great distances by reading their names or

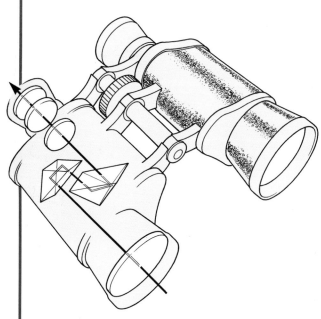

Above A section through a pair of modern prismatic binoculars shows the passage of light via two multi-faced prisms and thence via magnifying and focusing lenses to the eye.

numbers. They also need binoculars that will function in poor light. Such a combination leads them to choose huge binoculars mounted on a stand or tripod since after all they do not have to carry them around. Seafarers have similar requirements, but do not enjoy the stability of a land-based watch point, so they choose less magnification and a lighter, easier-to-steady binocular that is also robust and waterproof. Horse-racing enthusiasts need binoculars similar to those required by birdwatchers, though they frequently seem to settle for an over-large pair with a case slung low on the hip like a gun-holster.

What magnification?

So birders have to make a compromise between magnification, light-gathering power, field of view, and weight. No one wants to carry an albatross around their neck all day, and heavy weights are very difficult to hold steady in use. A magnification between seven or 10 times is perfect for birding and should be chosen by all. The field of view and light gathering power of a binocular depend on the size of the object lens – the big one at the front. By convention these lenses are described by their

measured diameter in millimeters, for example 25, 30, 40, or 50. A full description of the specification of a pair of binoculars is expressed as the magnification (say 10) times (x) the diameter of the object lens (say 40). Thus my own binoculars are 10x40, and very good they are too.

The basic description is, however, only the start, for binoculars vary enormously in weight, field of view, sharpness, twilight power, and so on. Modern "roof prism" binoculars have no external moving parts and are also relatively dustproof and damp-resistant. They have a central focusing knob and a variable adjustment on one eyepiece that enables the individual to compensate for differences between the eyes. Some makes are expensive, some relatively cheap; frankly you get what you pay for. If there is any chance of your becoming a keen birder, buy the best you can afford in the range 8 to 10 magnification; 35 to 50 diameter object lens.

The best advice of all is to see what most other birders are using and give them a trial.

Right Though binoculars are the essential everyday tool of bird watchers throughout the world, there are a few sites, such as this colony of sooty terns, where they may actually be a hindrance.

TELESCOPES AND TRIPODS

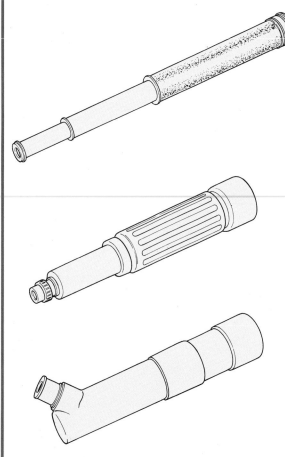

draw, brass contraptions better suited to the battlefields of World War I than the modern, high-speed campaign favored by contemporary birders. Today's optics are light, easy to use, and highly efficient. However, as with binoculars, it is all a matter of compromise.

Below The telescope and tripod combination is becoming a standard piece of birding equipment, especially when watching birds along shorelines and other open areas.

Right Though the telescope and tripod are heavy and cumbersome to carry, they have an immense advantage over the hand-held telescope in use.

Three quite different telescopes show the range of instruments now available. **Top** An old fashioned, multiple-draw 'scope that is focused by pushing and pulling. **Middle** A modern single-draw 'scope focused by turning the eye-piece. **Bottom** A prismatic scope is much shorter and therefore much easier to handle.

Armed with a decent pair of binoculars you can watch birds anywhere in the world, but it soon becomes obvious that many birds can be picked out, but not identified, at greater distances. Today more and more birders are making use of telescopes which can be clamped to a car window or a blind, or mounted on a tripod. Interestingly, most of these instruments were designed for use by target shooters, but manufacturers are at last getting around to the idea that birdwatchers are a distinct market with their own criteria of judgment.

Not so long ago telescopes were heavy, multi-

Lenses and magnification
Magnifications vary from about x20 to x40 and beyond, and there are lenses that will zoom through a huge range of magnifications. These zoom scopes would seem to be perfect for birding, but nothing is for free. The very structure of a zoom lens means that one loses on the field of view, making it less easy to pick out a particular bird. In fact, most zooms can "see" about 60-70 percent of the area that can be seen with a fixed magnification of the same power. For most purposes a magnification of x20 to x25 is perfectly satisfactory – though it is a good idea to choose a scope on which the eyepieces (that is, the magnification) can easily be interchanged. The addition of a x30 or x35 eyepiece may just come in handy on the odd occasion.

As with binoculars, the other half of the equation is the diameter of the object lens measured in millimeters. The larger the diameter the more light is gathered, but the heavier the 'scope. The usual compromise is between a 50mm and 70mm object lens, with a tendency toward the larger diameters with new lightweight glass.

At the time of writing the "draw" telescope is on the wane and even single draw 'scopes are probably doomed to extinction. The 'scope of the

moment and, dare I say the future, is a prismatic with no moving external parts. As ever, you pay for what you get.

In the U.S. there is a vogue for low-powered astronomical reflecting telescopes that work on the same principle as a photographer's mirror lens. The performance and cost are often staggering but, along with many others, I have reservations about the robustness of these instruments.

The prismatic 'scope

Settling then on a modern prismatic 'scope with a x20 to x25 fixed magnification and a 60mm or 70mm object lens the choice remains between a "straight-through" or "angled" eyepiece. For many years I have been a confirmed devotee of the latter. Angled eyepieces have several major advantages. Firstly, they can comfortably be used on lower tripods. This means that they suffer less buffeting in the wind and that you can "pass" a bird to a companion through the 'scope without having to lower (or raise) the height of the instrument. The latter is particularly important if, like me, you spend much of your time showing birds to others. Second, it is much easier to watch a bird higher than oneself, such as a bird of prey in the air, without being either a contortionist or a limbo dancer. The only possible disadvantage is that it may be more difficult to locate a particular bird with an angled eyepiece, though I have never found this a problem myself and it may be only the opinion of those who do not use the "angled" eyepiece on a regular basis.

Naked eye

8 × binoculars

20 × telescope

Magnification
The effect of using modern binoculars (× 8 magnification) and a telescope (× 20 magnification). Note that although the bird appears closer the greater the magnification used, the field of view decreases dramatically making the bird progressively more difficult to find.

THE COMPLETE BIRDER

With a good pair of binoculars, and a telescope, it is reasonable to suppose that one is prepared for birding at all seasons. There has, however, been something of a revolution in birding equipment over the past 20-odd years, so that the modern enthusiast often has as much equipment as the dedicated fly fisherman. Much of such paraphernalia is of marginal value and fit for only occasional use. And much has more to do with one-upmanship than with practical birding.

The equipment boom can be divided into a number of categories the most significant of which is optical gear. To add a second pair of binoculars to one's armory seems sensible. Even the most expensive glasses can break down and cause missed opportunities and embarrassment, so a spare can avoid disappointment. A spare, higher powered pair of binoculars can also be very useful, on long sea-watches, for example. A magnification of x15 or even x20 may be impossible to carry around the neck all day, but is marvelous for watching autumn skuas or summer shearwaters. They are certainly much kinder on the eyes than hours peering through a telescope.

The advantages of a tripod

Telescopes do, however, have great versatility, particularly when mounted on a tripod. Indeed the telescope and tripod combination is now the second favorite piece of equipment after binoculars. For preference the tripod should be as light as possible, but also as sturdy and firm as possible. This contradiction leads inevitably to yet another compromise: the best telescope support is immovable, yet mobility is of the essence. Several makes are favored by birders, but an easy action and light weight are features that should govern the choice.

Telescope-tripod combinations are awkward to carry about and shoulder straps are essential. These can be attached to the tripod by using the large split rings designed to hold keys. The broader the strap the more comfortable the contraption is to carry. Before leaving telescopes it is worth mentioning again the interchangeable eyepieces that are available for several of the better known and more favored makes. Most birders prefer x20 to x25 fixed magnifications and

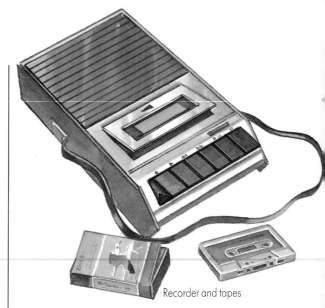

Recorder and tapes

Sun hat

Hiking boots

many carry a spare, say x30 to x35, for those occasions when a little more power can convert a "probable" into a "definite" species sighting. Personally, I don't; but I am quite happy to borrow a 'scope from someone else who is prepared to lug around a lot of extra gear.

Protecting optical equipment against accidental damage seems sensible, but is generally approached in a somewhat cavalier fashion by most birders. To be seen with a manufacturer's carrying case over the shoulder is positively out! Whereas to have a rain guard that slips up and down the binocular strap is definitely in! However, this should slip up and down one strap, not two as the designer intended. Telescopes are generally

Binoculars

Telescope

better treated and many birders bind them with protective tape to prevent chipping of the enamel. Some Scandinavian birders cover their 'scopes with expanded polystyrene, creating a huge white sausage with holes for the working parts. This may be going a little too far, but it is a good idea to tape any joints through which water could enter the optical system.

Clothing

The "one-upmanship" of birding depends largely on a set of visual signals. These are all designed to show that one is a "real" birder, who travels the world, sees rare birds virtually daily and, above all, spends long hours "in the field." Binoculars must,

therefore, look used not new. Telescopes should be taped, though the tape should be worn. Such "distressing" is even applied to clothing. Bruce Campbell, the well-known writer and bird watcher, once suggested that throwing a new jacket into a chicken run was the best method of "distressing," but contemporary chickens seldom have such freedom.

Birders today use tape recorders, cassettes of bird songs, and a variety of distressed clothing for various weather situations, as well as tripods, telescopes, interchangeable eyepieces, and binoculars. They also tend to sport hats, usually covered with "bird" badges, and cars similarly adorned with "bird" stickers.

61

IDENTIFICATION

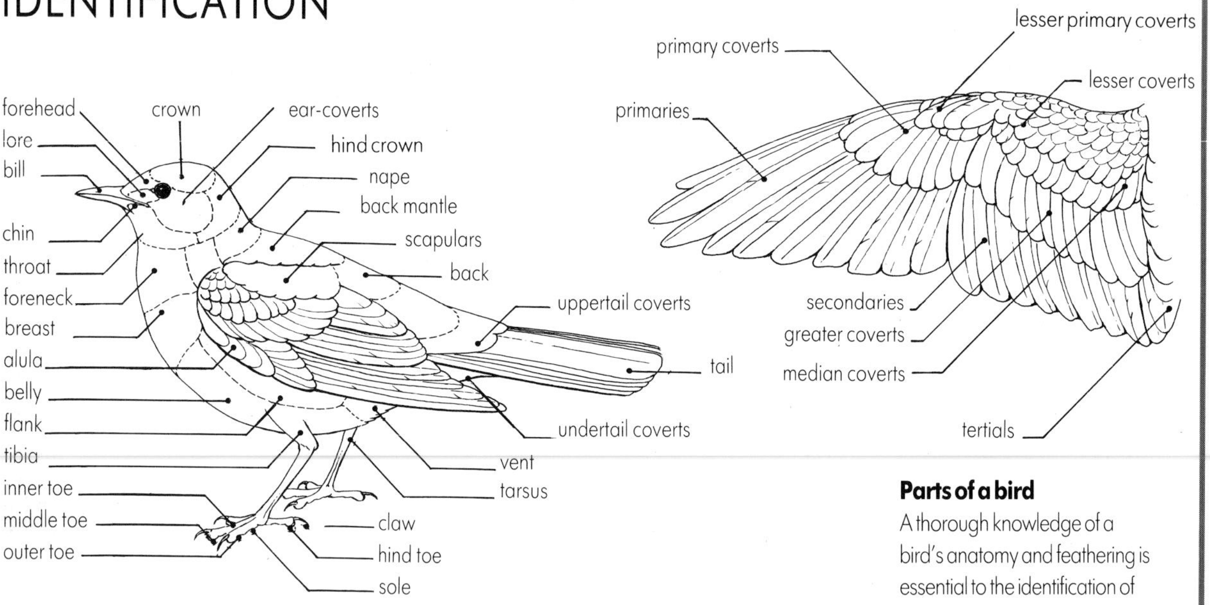
forehead
lore
bill
crown
ear-coverts
hind crown
nape
back mantle
scapulars
back
chin
throat
foreneck
breast
alula
belly
flank
tibia
inner toe
middle toe
outer toe
uppertail coverts
tail
undertail coverts
vent
tarsus
claw
hind toe
sole
primary coverts
lesser primary coverts
lesser coverts
primaries
secondaries
greater coverts
median coverts
tertials

Parts of a bird
A thorough knowledge of a bird's anatomy and feathering is essential to the identification of many species.

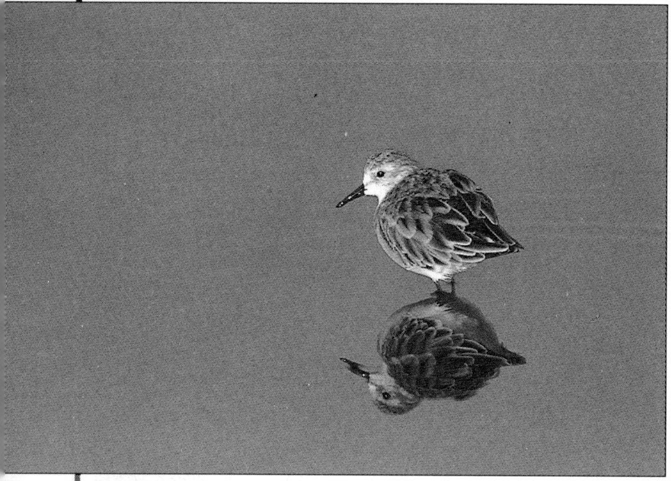

Above Small shorebirds known as peeps in the U.S. and "stints" in Europe pose identification problems of the highest order, particularly when, like this little peep, they are in nondescript winter plumage.

The primary skill of birding is being able to put a name to every bird seen, even with the briefest and most distant of views, and many highly skilled watchers never get beyond this fascinating aspect of birds. Much of this ability relies on sound knowledge of a bird's anatomy and, in particular, on knowing which feathers are which. Learning names like tertials, scapulars, and median coverts may seem a bit like college-level biology, but it is thoroughly worthwhile and very rewarding. What's more, it is essential if a bird is to be accurately described.

Knowledge; knowing what is where at what times; knowing how various birds spend their day and night; knowing how birds behave; how they fly; whether or not they form flocks, and of what size; in fact knowing everything one can about birds is the major aid to identification. In a nutshell, it comes down to knowing what to expect and then knowing what to look for.

Identification points

Though most identification books tend to concentrate on the color of birds, with abbreviated plumage descriptions, most good birders have a sound knowledge of structure to act as a basis of identification. In fact, even beginners have an elementary knowledge and can identify a duck as a duck, a wader as a wader, a hawk as a hawk, and so on. Learning the characteristics of other groups, picking out a warbler, or a thrush, a loon, or a grebe as such, is not difficult, and is largely a matter of structure.

Other structural features may be a little more elusive, but still provide the most important clues to a bird's identity. The shape of the wings can be particularly useful. Are they long and narrow like a swift's? Are they short and rounded like a grouse's? Are the wings broad and square like a turkey vulture's, or are they rounded like a sharp-shinned hawk's? Are they sharply angled like a swallow's, or are they straight like a shearwater's? With all birds wing shape is important, but with birds of prey shape may be the most significant feature of all as many such birds are uniformly dark brown.

Tail shape too can be vitally important. Some birds have long pointed tails like pintail and long-tailed jaegers. Others have deeply forked tails like the barn swallow. All other birds fit somewhere into this spectrum of tail shape with wedge-shaped, rounded, square, or notched tails. But shape and length are not the same and the proportionate length of a bird's tail may be one of its most significant field marks.

Similarly, not only do the shape and structure of a bird's bill tell us much about its lifestyle, but they can also prove an invaluable aid to identification. Length of bill may be difficult for the beginner to express, for terms like "long" and "very long" are more or less meaningless. So birders compare the length of a bird's bill to the distance between the eye and the base of the bill (the loral distance). Or, for really long bills, to the length of the head through the eye. We say, for example, that an Eastern bluebird has a bill about the same length as the loral distance, and that a godwit has a bill about 2 1/2 times the length of the head. Similarly, the length of tail can, in flying and particularly soaring birds, be compared with the width of the wing. In many eagles these are about equal, though a slightly longer tail can change one species of eagle into another. A harrier, on the other hand, has a tail that is almost twice as long as the width of the wing.

Size too can be a very useful identification tool, though it should be approached with great caution. An unknown bird should always be compared with that of a species that is well known and we say that a bird is "sparrow-sized," "starling-sized," and so on. Better still is to relate a bird with a known species with which it keeps company and can be directly compared. Even here, however, one should be careful of the effect of binoculars which can make a bird further away appear larger than a closer bird. If you find this hard to believe, try looking through binoculars along a railway line and see how the parallel lines appear to diverge rather than come together in true perspective.

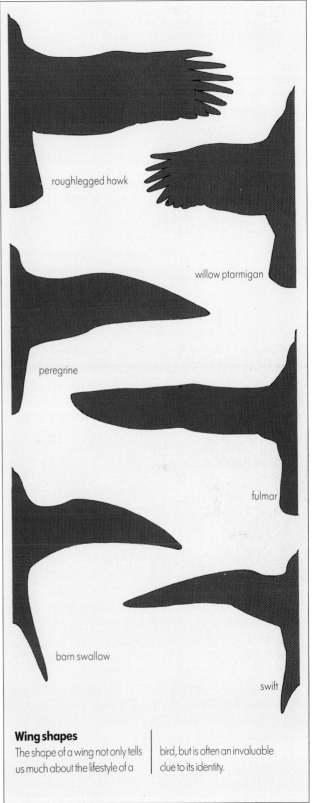

roughlegged hawk

willow ptarmigan

peregrine

fulmar

barn swallow

swift

Wing shapes
The shape of a wing not only tells us much about the lifestyle of a bird, but is often an invaluable clue to its identity.

63

FIELD GUIDES

The main aid to identification, even in an increasingly technological age, is the field guide, a book that illustrates and describes all the different birds one is ever likely to see in a particular area or region. Even 30 years ago, color printing was something of a novelty: today we take it for granted and it is unthinkable that a modern field guide would not be illustrated in color throughout. This facility means that everything we need to name a bird can be grouped together rather than scattered somewhat randomly over several pages. Color printing is, however, still expensive and inevitably the publishers of any field guide have to make a choice between producing an expensive book that deals in depth with a small area for a small market, or a cheap one covering a much larger area and for a correspondingly larger market. In Europe this tends to mean that field guides cover the whole of the continent rather than just an individual country. In America they cover the whole country, rather than any individual state. Only by printing tens of thousands can the high cost of color illustration be offset to produce a cheap book.

What to look for

The essential ingredients of any field guide are color illustrations of the bird; a descriptive text that complements rather than repeats information supplied by the illustration; and maps showing where the bird is found. A good field guide will, additionally, illustrate every major plumage of every bird, including sex and age differences as well as illustrating both the upper and under surfaces of the bird in flight if that is how they are often seen. The text will describe each of these plumages, as well as picking out where and when the bird is found, together with some idea of its abundance and "visibility." The maps will show not only where a bird is found in summer when it is breeding, but also where it winters and areas that it passes through on migration.

As we enter the 1990s, there is a wide range of field guides available which, to a varying extent, meet these criteria. They cover various parts of the world, though inevitably the widest choice covers those areas - Europe and North America - that have the highest populations of birdwatchers.

These areas have several excellent field guides from which to choose and, although we all have our favorites, there is a great deal to be said for having several in the field. The Peterson field guides (eastern and western editions) are the traditional favorites, constantly revised since their publication more than 50 years ago. Among the newer guides, two excellent choices are *Field Guide to Birds of North America*, published by the National Geographic Society, and *Birds of North America*, a Golden Field Guide. The Audubon Society *Master Guide to Birding*, a three-volume soft-bound set, is too bulky to be a true field guide, but is an excellent reference.

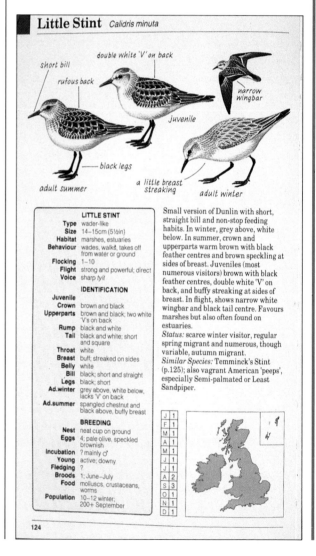

Left This guide shows each bird in both summer and winter plumage, as well as in flight, making it easier to make a correct identification.
(Courtesy Kingfisher Books)

Above A good field guide is essential for anyone interested in birds. There is a vast selection available, ranging from those which cover specific areas to the more general.

Elsewhere in the world, field guides vary from the excellent to the pathetic, with some countries being well served and others without a single book to aid the traveler. Visiting even well-known bird areas such as East Africa, India, or parts of South America can, therefore, be a very frustrating experience. The main problem is the cost of completeness, for with an avifauna of anything over 500 species the expense involved in producing a full-color guide cannot be recouped from sales to visitors alone.

One of my colleagues has overcome the problem by buying two copies of every field guide available for anywhere in the world, plus two copies of any other well-illustrated bird book. He then sets to cutting out the individual birds and pasting them into scrap books for his journeys. This may work out expensive, and I personally cannot bring myself to cut up books! Nevertheless, it is one answer to the problem.

Field guides are not only essential for identifying birds in the field, they are also excellent bedside books. When beginning to study birds, or when planning a trip to a new area or region, taking the field guide to bed and poring over the possibilities is a splendid way of getting to know the birds that will be new to you. Slowly but surely, you get to know what to look for in an individual species and, most importantly, you find out in advance the often subtle differences between closely allied and similar species. Anyone, for instance, intending to visit Africa should work thoroughly through the various species of weaver. Many are golden with black heads in the male and only by knowing the minor distinctions will most of these birds be identified. Anyway, as field guides lack any sense of plot, they are generally better at sending one to sleep than spy stories.

FIELD NOTEBOOK

Opinion is divided as to whether or not one should carry a field guide in the field and whether to consult it when faced with an unidentified bird. In general it is thought that with the bird before you, it is likely that the book will suggest identification features that you will then manage to find, thus hoodwinking yourself into seeing what you want to see rather than what is actually there. Purists, therefore, feel that taking a good set of field notes should precede book consultation.

Writing a description of an unknown bird requires practice and it is surely better to practice and perfect the technique in advance than to wait for a rarity to turn up and then find that you lack the skill to describe it. Even a humble garden bird makes a perfect subject for a field description.

Compiling a description

The description should start with the circumstances. Where and when was the bird seen, what was it doing, what other birds were associating with it, what size was it, and what were your overall impressions? Next describe, part by part, the upperparts of the bird starting at the forehead and progressing via crown, nape, mantle, wings, and rump to the tail. Follow with a similar description of the underparts from chin, throat,

Below Identifying ducks in flight depends on picking out the salient features, often marks on the upper wing. Knowing what to look for on each species is essential.

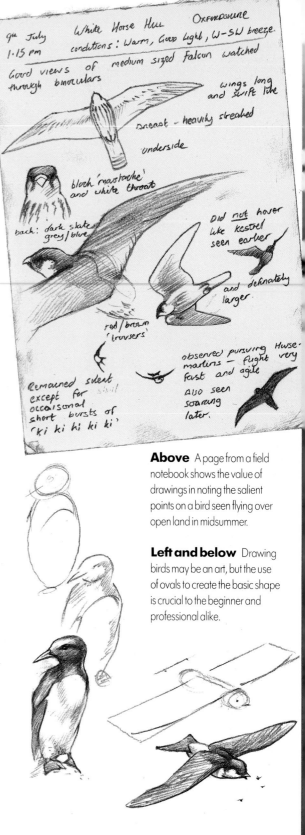

Above A page from a field notebook shows the value of drawings in noting the salient points on a bird seen flying over open land in midsummer.

Left and below Drawing birds may be an art, but the use of ovals to create the basic shape is crucial to the beginner and professional alike.

neck, breast, belly to undertail. Learn the names of the different feather groups, especially those of the wing, both when folded and when spread, and use the names in your description. Next describe the bill and legs, for some reason often called the birds "soft parts"! Finally if, while identifying the bird, you considered other species, put them down together with the reasons you had for eliminating them. At the end state, in percentage terms, how sure you are of the identification. You may care to make a rough drawing of the bird, using arrows and notes to pick out the diagnostic points, and many birders make such sketches as a form of shorthand prior to a detailed description. Drawings are also very useful for picking out undescribed identification features to add to your knowledge of what to look for the next time the bird is encountered. At first, these drawings will be pretty crude, but it is surprising how quickly one can progress. Remember that birds are just two ovals joined together in varying relationships.

Permanent records

The other major use of a field notebook is to keep a record of what you have seen. In some cases this

Above left The European spotted flycatcher is a small brown bird virtually devoid of field marks. It does, however, have the characteristic habit of flying out from a prominent perch, catching a fly and returning to the same place.

Above Correctly identifying American flycatchers is often a matter of making careful field notes and checking the crucial points with handbooks. This great crested flycatcher can be confused with three other very similar species.

is no more than a list of birds together with the number of individuals. Some birders produce more detail and some virtual works of art. Huge amounts of data stored away in birders notebooks tends to remain uncollated, and thus wasted. Modern personal computers have an enormous capacity for storing and sorting information and thus are ideal for record keeping.

At the end of the year, it is a great pleasure to thumb through the pages as a sort of vicarious reliving of the birding experiences enjoyed. It is also a good time to pick out the significant from the mundane and write up the records for the local county or state bird report. There are a host of such publications all waiting to receive reports of every bird that occurs in their area.

PHOTOGRAPHY

Photographing birds is worth a book in its own right and many have been written over the years. Most are full of good advice, describe the various techniques and equipment, and are written by experts. However, in my own mind at least, I divide bird photographers into two distinct types: there are bird watchers who take up photography, and photographers who choose birds as a subject.

Photography is an art form and great photographs are taken with care, thought, and an artistic eye. Most professional photographers employ an assistant who actually releases the shutter to expose the film. Bird photography does not quite fit such a process, though the nearer it comes to doing so the better the finished result.

The choice of film

The first thing to do if you want to photograph birds is to decide how you want to use the finished results. In its simplest terms this may involve a choice between film for producing prints to illustrate a diary, or film to produce a slide show for your friends or a local bird-watching group. You may, however, harbor dreams of selling your work; if so then great care must be taken in selecting the right film. All color photographs that appear in magazines and books are taken from transparencies (slides). They are chosen by professional editors and designers not only on the basis of content, but also on technical merit. As a general rule, a slow, high-resolution, low-grain film produces a transparency that will reproduce in print much, much better than a faster film. So the film choice is limited and many transparencies must be discarded because of out-of-focus

Below Photographing small, fast moving birds in dark woodland conditions poses a problem because of the reduced light available. Many photographers overcome this by using an electronic flash. This hide is set at a pied flycatcher's nest.

Above To reach the nest of these golden orioles a tower hide was needed, and the photographer was rewarded with this delightful portrait of the pair together with their young.

Right Hides often need to be raised from the ground so that tree nesting birds can be satisfactorily photographed. Not all are as conveniently situated as this mistle thrush nest in a back yard.

subjects, or subjects that move. Modern, fast color films that can be up-rated may produce lovely slides to show your friends, but they will not stand up to the test of reproduction.

Camera selection

Choice of film is then the first crucial element and its importance cannot be overstated. After that, all you need is a camera and you are into bird photography. Surprisingly, the modern single-lens reflex camera using 35mm film is used by both amateurs and professionals alike. These cameras are the most popular and widely available and are within the reach of even the shallowest of pockets. The only real essential for bird photography is the facility to change lenses.

To many, the idea of bird photography immediately conjures up a vision of huge telephoto lenses, but just as one should always use the slowest film possible, so too should one use the shortest lens that will do the job. With many small birds this will nevertheless be a telephoto of, say, 135mm, but the secret is to get closer to the subject rather than use a greater magnification.

Having put the, dare I say, professional standpoint, it is reassuring to know that there are hosts of people taking splendid bird photographs on fast film with huge lenses. Most use 35mm SLR

cameras with lenses varying from 300mm to 1000mm in focal length. They use motor drives, automatic exposure systems and, sometimes, self-focusing devices. They take birds in bushes, on mud-flats, and along the shoreline, both flying and standing still. Many of these action shots are obtained by stalking birds, using a shoulder holster for steadiness.

Others are more careful and more traditional. They set up blinds at nests so that they can work closer to the birds and spend hours in preparation to obtain the perfect portrait. There is, however, a definite move away from nest photography and toward using the same "controlled" techniques on

other aspects of birds' lives. Birds come to feed and to drink in much the same way as they come to their nests and these activities offer similar, more varied and interesting, opportunities for photography.

There is, I am pleased to say, also a more careful and artistic element entering into bird photography. Birds are now being photographed in their habitats, showing them more as we see them while birding. Over and over again such photographs are being chosen as winners in competitions so that expensive and complex electronic equipment is taking second place to the eye of the photographer. To anyone starting out in the art, this is the most encouraging sign of all.

RECORDING

The modern, portable tape recorder has revolutionized our approach to bird songs and calls. Thirty or so years ago, recordings were cut directly onto disks and the word "mobile" was used to indicate that a caravan, linked to a main electricity supply, could be used as a transportable studio. Tape recorders changed that and modern cassette recorders have made it even easier to get out and about recording the sounds of birds wherever they can be heard.

The art of making recordings of bird song centers on isolating what has to be recorded from the general background of other noise. It is not sufficient to seek a quiet spot away from traffic, turn on the recorder and tape what can be heard. The result will be a quite unacceptable cacophony

Left Wildlife film-maker Liz Bomford uses a parabolic reflector to obtain the required quality of sound for her television documentaries.

Above Portable tape recorders, combined with a parabolic reflector to focus the sound of a singing bird, enable even flying birds to be recorded.

of mixed-up sounds. Wind, even light wind, is the major enemy, but even on a still summer's morning miles away from roads, farmyards, chain saws, and aircraft, the countryside is still not quiet.

Microphones and reflectors

Isolation of the individual call requires a directional microphone similar to that used by outside broadcasters, film makers, and at sporting events. Mostly these professionals use a sausage-shaped object, often covered with fur and resembling a shaggy dog. This "sausage" contains a microphone and picks up sound mainly from the direction in which it is pointed. It is a robust, easy to handle form of directional microphone. It does, however, function best at close range. The alternative, and the solution chosen by most bird song recordists, is the parabolic reflector. This is a circular dish that concentrates the sound from a relatively narrow field to a single focus where the microphone is located. It thus works in the same way as a television dish for receiving signals from satellites.

Parabolic reflectors do not need to be huge and heavy and are available commercially at a reasonable price from several specialist manufacturers. Armed with one and a good quality cassette or tape recorder, anyone can record the songs and calls of birds in the wild.

Some recordists simply wander through likely spots recording as they go. Others treat recording as they would photography, and make elaborate preparations to record the particular species selected. This may involve placing the reflector close to a well-used song post with a lead to the recorder set up some distance away. In general, the longer the lead the less quality is obtained in the recording, so some recordists use blinds just like photographers to obtain a really close approach. A few purists actually prefer to abandon the reflector and obtain a more "natural" recording by placing their microphone right at the bird's song post.

Playing a recording

As with photography, there are several books available on making recordings of wildlife sounds, but for most of us the tape recorder serves quite a different purpose. Even a cheap cassette recorder with a built-in microphone will pick up a bird singing at relatively close range. If the recording is then played back to the bird, individuals of many species will approach to investigate, thus giving us an opportunity to view them openly and at close range. There are also commercially available recordings offering a huge range of bird songs and calls that can be played with similar effect. Calling birds up, in this way, can produce views of skulking birds and help to locate birds that have not been seen (or heard) at all. But using recordings in this way is not without its dangers.

Most birds that react to recordings will be on their breeding grounds, defending a territory. Investigating intruders and singing to defend their domain is a full-time job without having to contend with a human, apparently with a competitor sitting in his hands. A quick play and a satisfactory response probably does no harm, but a lengthy intrusion and particularly repeated intrusions may well interfere with the bird's breeding activities. So, as ever, moderation and thought for the bird are the keys. Some ornithologists and conservationists believe that one should never play recordings to birds on territory – certainly rare birds should never be disturbed.

Right A casual, rather opportunistic approach to recording bird songs can be enjoyed by anyone, even on an ordinary walk through woodland.

LISTING

Right A typical gathering of twitchers on the Isles of Scilly, England, in fall. Similar, if smaller, groups of listers can be found on the island of Attu, the farthest west of the Aleutian Islands, Alaska.

When we start birding, new birds come thick and fast and most of us, quite naturally, keep a list of all the different birds we see. Gradually, new birds begin to become harder to discover, though conversely the thrill of seeing one increases. After a relatively short time, the new bird for our list becomes exceptional and we may lose interest in listing altogether. We may become interested in banding, photography, the study of a common species, or in traveling to new areas to find different birds.

For many birders, however, the lure of listing hangs on. Most birders keep a "life list," simply a record of all the species they have seen, usually with a note as to date and place. Others take the concept a step beyond – keeping country lists, state lists, county lists, even property lists; lists of birds seen on particular journeys or in particular refuges; lists of breeding birds or lists of migrants seen passing on a single day; lists that span a year, a month, or a day, such as the annual "Big Day" outings in many parts of the U.S. and Canada.

Rarities or "accidentals" (birds that are common elsewhere but have strayed off-course) are the bread-and-butter of many listers. Some with time and money to spare will even keep a travel bag constantly at the ready, so they can dash off at a moment's notice should they hear of a rare bird somewhere. The use of telephone answering machines has been a boon to listers (and others who enjoy chasing accidentals). Many birding organizations have used them to set up hotlines that are updated regularly to give callers the latest information on sightings. For the more serious, there are even networks one can join that promise an instant alert whenever rare birds are located anywhere in North America.

Merits and demerits

Opinions on the merits of listing (often known as "ticking") are sharply divided among birders. Hard-core listers extol the excitement and competitive feeling that they get from this pastime within a pastime. Others view the whole exercise with distaste, believing that a concentration on listing robs birding of its introspective nature and reduces a bird to nothing more than a mark on a piece of paper.

The latest wrinkle in listing is birding-for-charity, exemplified by the World Series of Birding, held each May in New Jersey. Organized by the New Jersey Audubon Society, the World Series pits teams sponsored by birding magazines, binocular companies, ornithological groups, and others against each other, each trying to compile the biggest list of birds in a single 24-hour period. Each team collects pledges on a per-bird basis, and routes are plotted out in extreme detail weeks in advance in order to ferret out the greatest number of species. The competition is, to say the least, tremendous.

The vast majority of birders happily occupy the middle ground, keeping a life list of some sort, enjoying new birds when they come across them, but getting the greatest satisfaction out of learning more about the common species they see regularly. Fortunately, birding is a pastime flexible enough to accommodate birders of every type.

Left Ross's gulls from the high Arctic pack ice occasionally stray southward to send twitchers and listers alike into a frenzy of hurriedly canceled appointments and mysteriously contracted minor illnesses.

Below A migrant melodious warbler in Britain often draws sizable groups of birders to ponder its identification, particularly the problem of distinguishing it from the similarly featureless icterine warbler.

THE BIRDER'S YEAR

Understanding the main ingredients of the birds' year is essential if one is to get the best out of one's birding. There is no point in looking for warblers in January, or wild geese in June; but a much deeper knowledge of the individual lives of a whole range of different species is required if the greatest enjoyment is to be obtained.

Spring arrivals

The second half of April and first half of May see the main arrival of birds that have wintered, often in foreign climes, well to the south. Many, like the swallows, warblers, and flycatchers, may breed nearby, but others may merely pass through on their way further north. This is an excellent time for identifying a wide range of birds, for most are in

Left Knot occur in summer plumage for a few brief weeks in mid-summer. The birder who misses this treat must wait a full year to try again.

Above Snowy owls breed in the far northern tundra zone. When the population of lemmings crashes, these birds come southward in search of alternative food.

their summer finery, in plumages that show them at their best. As so many birds are singing lustily, this is also the best time to locate the scarcer or more secretive birds.

In general, spring is not as good as autumn for producing rarities, but a phenomenon called "overshooting" regularly produces birds that breed further south. In fine, warm weather birds often fly straight over their normal breeding zones to appear in areas often hundreds, and sometimes more than a thousand, miles to the north. That is they "overshoot" their range.

Soon things begin to quiet down as residents and summer visitors alike settle down to rear their young. A few particularly high-Arctic birds may

still pass through during June on their way to the tundra wastes, which remain frozen solid until even early July. But June is the month for watching breeding birds, enjoying bird song, and seeking better views of the more elusive species.

By mid-July the countryside is full of young birds. By the end of the month waders are returning from further north as the vanguard of the masses that will pass through in August. For some peculiar reason August has gained a reputation as a poor month for birds. Yet nothing could be further from the truth. With a summer's worth of fledglings filling the woods and fields, there are more individual birds than at any other time of the year. The problem is that, with the breeding season finished (except for some perennially late nesters like the American goldfinch and the cedar waxwing), the males have stopped singing - and therefore many birders have trouble seeing any birds.

Migrations in the autumn

The autumn migration is already picking up steam in August. The shorebird migration peaks in mid-month as the sandpipers and plovers pour south from the Arctic on their way to South America. The first waves of songbirds appear in the middle latitudes, a tide that builds toward the end of the month. Likewise, bald eagles and ospreys that have summered in the north start southward along the mountains and coast, a presage of the main fall flights. September, however, is when the full rush of migration arrives. Most insectivorous songbirds are on the move, although not always in large, obvious flocks. Swallows, especially tree swallows, may mass in incredible numbers along their coastal migration routes, sometimes by the millions. Equally impressive is the September migration of broad-winged hawks in the East and Swainson's hawks in the West; at times, more than 25,000 broad-wings have been seen in a single day at Northeastern hawk lookouts, and totals of more than 60,000 are not unknown in Texas, a bottleneck on the birds' migration to South America.

The birder's winter
October brings many of the larger birds – Canada and snow geese down from the Arctic, seabirds off the coasts, a whole range of raptors along the mountain ridges and seashore. Most of the summer resident songbirds will be gone, and the first of the winter visitors – juncos, white-throated sparrows, evening grosbeaks, and finches – will be

arriving. Fall is also an excellent time for finding rarities and accidentals; out of the millions of birds on the move, there are always a few with a misguided sense of direction, or which have the misfortune to be blown off-course.

By November, most of the migrants have arrived at their destinations. A few laggards, like red-tailed hawks, golden eagles, and rough-legged hawks, are still on the move, however. Others arrive in the U.S. with no particular regularity. These are the "irruptive" species, usually northern raptors like snowy owls and goshawks that normally spend the winter on their breeding grounds. But food shortages, caused by population cycles among their prey species will force them to wander south every few years. Owls, in particular, follow no special routes, nor do they head for a traditional wintering ground. They simply drift until they find a suitable subsitute for the tundra.

Midwinter, from December through early February, is the bleakest time of the birder's year in the northern latitudes. The selection of both species and individuals is lowest, except at coastal refuges that play host to large numbers of waterfowl. But the first hint of spring is enough to send the birds north again. The ducks and geese are usually in the vanguard, probing north in the wake of receding ice, filling the air with their calls.

Left To see spotted redshanks in their black finery is possible in June and July. To see them perched in trees requires a special journey to their northern breeding grounds.

Below Rough-legged buzzards leave their northern breeding grounds on an irregular basis, depending on the population of small mammals.

URBAN BIRDING: CITIES

An urban birder can be forgiven for looking beyond the city limits for better sport; after all, the country holds by far the greater variety of birds. But don't write off the urban landscape entirely. A surprising number of birds have adapted to life in the concrete jungle. The most successful are imports from Europe, which have since grown to pestiferous numbers.

Imports from Europe

In the mid-1800s, cages full of live house (or English) sparrows were shipped across the Atlantic at the request of Americans who thought this lively little member of the weaver family would be a good addition to North America's avifauna; others hoped it would help control such agricultural pests as cutworm. House sparrows fanned out across the U.S. and Canada like wildfire, at home almost wherever people lived, but especially in cities and on farms.

Starlings followed much the same track. A group called the American Acclimatization Society, its lofty goal to establish in the U.S. every species of bird mentioned in the works of Shakespeare, released a batch in Central Park in 1890; within the century the species had spread from Florida to southcentral Alaska.

The North American origins of the pigeon, or

rock dove as the wild strain is usually called, cannot be as precisely traced. A favored food bird since the days of the Greeks, it was brought to the New World early on, living in plantation dovecotes. With the rise of cities, the rock dove reverted to a feral lifestyle, living among the towering buildings that so resemble the cliffs the species still inhabits along the Mediterranean.

Adapting to urban life

Cities are not solely the haunts of alien species,

Left The peregrine falcon breeds on mountain buttresses and sea-facing cliffs over much of its range. It does, however, frequently resort to city tower blocks in winter and will even stay on and nest.

Below Central Park, New York, offers a green oasis in a sea of concrete that draws migrant birds and their watchers throughout periods of passage.

however. Many native North American birds have also adapted to urban habitats. The common nighthawk once nested on graveled river beds and forest fire burns, but it quickly switched to cities, where flat-topped buildings, covered with tar and a layer of crushed stone, suited it to perfection. An insect-eater that is a nightjar rather than a true hawk, the nighthawk is a cryptically colored bird

Left Although basically a bird of open ground where there is only scanty cover, the nighthawk has taken to nesting on flat roofs of factories in several parts of North America.

that takes to the air as the summer evening approaches. It flies gracefully on long, tapered wings, swooping after airborne insects. Its eggs are laid directly on the rooftop without any pretense of a nest. Among nighthawks, the biggest problem during incubation is preventing them from frying. The female must shade the eggs through the long, brutal day, when temperatures may reach 140°F on the oven-like roof.

The chimney swift is another confirmed urbanite, and one that (to judge from its name) apparently took to human habitation early on in its U.S. career. North America's other three species of swifts, the white-throated, Vaux's, and the black, nest in hollow trees or crevices in cliffs. A few chimney swifts still do, but most seek out unused chimneys, silos, or other vertical, enclosed spaces. Here they build a semi-circular nest of small twigs, glued to the wall with their thick, sticky saliva. Chimneys also serve as communal roosts during the late summer and in the migration, and the sight of thousands of swifts like a gray cloud funneling into a tiny opening is arresting.

City predators

Few birds have taken to city life with the gusto of nighthawks and swifts, but where there are tree-lined streets and yards, it is usually possible to find American robins, grackles, mockingbirds, and a selection of other small songbirds. The presence of prey attracts a few predators as well. Especially in winter, young sharp-shinned and Cooper's hawks may move into the less hectic parts of town, stalking house sparrows and starlings. Great horned and screech owls occur in cities more frequently than most people suspect, feeding on the rich supply of rodents. The monarch of city birds, however, is the peregrine falcon, the fastest creature on earth. A wilderness bird for the most part, peregrines were found in small numbers in metropolitan areas prior to World War II, nesting on skyscrapers and feeding mostly on pigeons. The introduction of DDT and other persistent pesticides nearly wiped the peregrine out across most of its U.S. range, and the city-dwelling falcons disappeared.

In the 1970s, however, The Peregrine Fund, based in Cornell University's Ornithology unit, began releasing captive-reared peregrine chicks, using an old falconry technique known as "hacking." The chicks were placed on towers and fed by hidden assistants until they could fly. They were then released, the feedings carrying on until they learned to care for themselves. Today, peregrines again breed in small numbers in several U.S. and Canadian cities.

SUBURBIA

Left House finches have spread rapidly in the eastern United States after being introduced from the west. The provision of feeders may well have played a part in their success.

Below The mourning dove is the most successful member of its family, being equally at home in farmyards, suburbs and cities. Its opportunism has enabled it to expand northward as far as Alaska.

Not quite city, not quite country, the suburbs have a little of both. Among the quiet yards and gardens, a surprising variety of wildlife makes its home – especially birds.

A new housing development, with its barren lawns and lack of trees, is scarcely inviting to birds, and only a few hardy varieties like American robins and starlings will be in any abundance. But within a few years, as trees are planted, hedges grow, and landscaping is accomplished, the quantity and quality of cover increases, and so do the birds. Within 20 or 30 years, shade trees will dominate a development, adding the songs of many woodland species to the dawn chorus.

Bold and adaptable

The robin is perhaps the quintessential suburban bird – an observation that obscures the fact that robins are among North America's most widespread birds, found from the Arctic to the desert. Yet a robin, head cocked low waiting for a worm, seems most at home on a carefully manicured suburban lawn. In much of the north the robins vanish for the winter, reappearing in early spring, as soon as the ground has thawed enough for them to hunt earthworms. Little time is lost in setting up territories; the males sing their musical cheery-up, cheerily, cheerio song incessantly, attracting mates and establishing the boundaries of their fiefdoms. Nest-building falls mostly to the female, who builds the cup out of grass stems plastered cleverly together with ample quantities of mud, usually siting the nest in the fork of a tree branch. The four or five eggs, tinted with the famous "robin's egg blue" in the female's oviduct, are incubated for about 14 days before they hatch. The chicks, naked and blind, nevertheless raise their open mouths to the sky whenever their parents arrive with food, clued by the jolt of the landing adult at the edge of the nest. The youngsters feather quickly, and are ready to leave the nest within two weeks of hatching. The adults may reuse the same nest, or – especially if the old cup is badly infested with parasites – may build a new one for their second and third broods.

Suburban birds tend to be the species that are bold and adaptable, the ones that can get along with humans and flourish in a much-changed environment. The blue jay is an excellent example of a bird that has learned to make the best of a changed situation. Once strictly forest birds, blue jays are now commonly found in suburbia, nesting in shade trees and feeding on the rich provender of gardens, feeding stations, and trash dumpsters, in addition to their normal diet of insects, fruits, berries, and small animals. Crows, likewise, have done well in suburbia, due in large part to their unrestricted diet and high intelligence.

Advantages of the "edge habitat"
The birds that have done best among the shopping malls, housing developments, and commercial sprawl are those that ordinarily live in what biologists refer to as "edge" habitat – not deep forest, not open field, but a brushy mix of the two. The regular interspersing of yards, hedges, vacant lots, and stream banks suit them to perfection, and so cardinals, mockingbirds, song sparrows, chipping sparrows, northern flickers, and common grackles are abundant. The mourning dove, once primarily a farmland bird, has taken to the suburbs in a big way. Two major reasons are credited as being responsible for the dove's continuing population explosion: birthrate and handouts. Mourning doves are prodigious breeders, nesting earlier and longer than almost any other songbird; in the Northeast, it is not unheard of for doves to be incubating eggs by the first week of April, or for newly fledged chicks to be seen in late September. A single pair may raise as many as six broods in a season, although three or four is more usual. Care of the eggs and young is shared between the sexes, with both the male and female producing "pigeon milk," a secretion of the crop lining, to feed the chicks. Once on their own, however, many doves turn to human help, especially in the winter. Mourning doves eagerly accept the seed scattered at feeding stations, and many scientists believe the free food is a major reason for the growing number (and expanding winter range) of this lovely bird.

Feeders may also be playing a role in the phenomenal expansion of the house finch in the eastern U.S. and Canada. Descended from several cagefuls of western house finches, and far from the arid scrub of its native West, the house finch has settled firmly into the suburbs, small towns, and farms, often at the expense of another non-native bird, the house sparrow. From its tiny beginnings, the range of the eastern population has grown to encompass most of the East; the western population is likewise expanding, and it seems likely that the Great Plains will eventually be bridged, resulting in coast-to-coast house finches.

Below The robin has learned to live happily alongside humans and has prospered as a result. It is now more abundant in the suburbs than in its natural woodland habitat.

DUMPS

Birding is usually a pursuit that pleases all the senses – the beauty of a bird's colors, the delicacy of its song, and the rich smells of a summer forest or a chilly, windswept beach.

Smell, however, is something best dispensed with at a dump, especially on a sultry summer day, when the stench can be gagging. So why bother hanging around a landfill at all? Because birds do, too. Not a wide variety, to be sure, but the number of individuals can be astonishing. Large municipal landfills may attract thousands of gulls, and lesser numbers of other kinds of birds.

Gulls and dumps

The king of the heap is the herring gull, the ubiquitous gull of the Northern Hemisphere. In many areas it has become, like the Norway rat, virtually a human commensal, living on the offal that society throws away. Certainly, herring gull populations have vastly outstripped the numbers that the land itself would naturally be able to

Above A wheeling mass of gulls completely ignores the compactor and its driver at this Florida dump. With care you will be able to identify several different gull species, as well as white ibis and cattle egret.

support, and have become almost wholly (and indirectly) reliant on people for their food.

It was not always this way. At the turn of the century, herring gulls in North America had been decimated for the feather trade and by eggers – so much so that fledgling conservation groups like the Audubon Society were worried about the species' imminent extinction. Protected by newly enacted laws, however, the gulls began a slow recovery. The pace quickened as more and more started using garbage dumps as feeding grounds. Their population growth became explosive.

Today, far from being pampered, threatened species, some gulls have become the focus of control campaigns designed to save other imperiled birds. Colonial seabirds on the

Northeast coast, especially terns, have suffered greatly from gull attacks in the breeding season. Many traditional colonies have vanished, and the roseate tern has been declared an endangered species, in large part because of gull depredations. In response, biologists have begun controversial programs to kill adult gulls and their eggs near tern colonies. The results have been encouraging, with some tern colonies rebounding quickly. While mass control of a wild animal is rarely a preference, those who argue against the gull campaign forget that herring gull populations are, in a sense, "subsidized" by mankind, and have grown far beyond the numbers that they would attain in a strictly natural situation.

Herring gulls are not the only gull species to use dumps, of course. In the East, great blacked-backed and (to a lesser extent) ring-billed gulls also forage in landfills, while glaucous-winged and western gulls do so in the West. Crows, starlings,

Left The black vulture is a regular inhabitant of garbage dumps, particularly where these are located near the coast. They are common birds over much of the southern and eastern U.S.

Above Although a fierce and impressive bird, the great horned owl is not averse to the easy pickings provided by the food that humans throw away.

and grackles, being adaptable omnivores, are usually quick to zero in on the abundant food, and in parts of the South, black vultures have been traditional dump visitors – although a decline in the number of large animal carcasses brought on by the rise of automobiles has decreased the dump's importance to this scavenger.

Attractive to predators

Because dumps are rich feeding grounds for rodents, it is little wonder that they attract many predators as well. Great horned owls are common nocturnal dump users. Probably the most adaptable raptor in the Western Hemisphere (they are found from Alaska to Tierra del Fuego), great horned owls will eat whatever fortune brings their

way. In densely forested regions, they may specialize in snatching songbirds from their nighttime roosts; in farm country, rabbits, opossums, and mice usually make up the bulk of their diet. They are unmatched as ratters, however, and where Norway rats are abundant they may eat little else. Thanks to a digestive quirk, it is easy for scientists to know what they eat. When an owl catches small prey like a mouse or rat, it eats the animal whole, or it is torn into several large, fairly complete chunks - fur, bones, feathers, and internal organs. Within eight to 12 hours the meat and other soft parts have been digested, leaving the bones and other indigestibles to be coughed up in a compact pellet. Because the owl usually regurgitates its pellets while on roost - thus making them easy to find - biologists have only to identify the remains to decipher the owl's feeding preferences.

SEWAGE PONDS

The corollary of birding at dumps is birding at sewage ponds, where treated wastewater is stored in open lagoons. Again, the setting may seem unsavory, but the birds – shorebirds in this case – make up for the unpleasantness.

With rampant development eating away steadily at North America's wetlands habitats, migrant shorebirds have faced increasing problems in finding suitable places to rest and feed on their journeys. From the air, a sewage pond presumably looks no different to a sandpiper than does a natural pond – and when they land, they find an abundance of invertebrate food.

Growth of interest

It has taken birders much longer to catch on than it did the birds themselves. Western bird-watchers have been the leaders in sewage pond birding, especially in California and the rest of the Pacific coast. While natural estuaries and mudflats generally hold the greatest number of shorebirds, sewage works have developed a reputation as the places to find rarities, and journals like *American Birds* routinely refer to well-known sewage ponds in the same breath as wildlife refuges.

The days are long past when a municipality could simply dump its sewage effluent into a convenient waterway. Today, expensive sewage treatment plants handle the task of purifying the fouled liquid, using a series of settling ponds, tanks, chemical and bacterial processes to remove

Below The greater yellowlegs is only one of a large number of waders that use sewage settling pools as handy migration stop-overs during their journeys through the interior of North America.

but to the plovers it is perfect, affording open spaces and insects. Likewise, large, freshly plowed farm fields attract ring-billed gulls, crows, songbirds – even great blue herons looking for worms, bugs and dazed mice.

One other "industrial" process also creates artificial marshland and is richer by virtue of being sited near the coast. Most great cities were originally ports, where goods could be bought and sold. Many are sited at the mouths of great rivers where ocean going ships could discharge their cargoes and where inland transport could continue by barge. New York and San Francisco, London and Rotterdam are all cases in point. As ships have grown bigger the survival of such ports has depended on deep channels to allow continued access to vessels drawing many fathoms. Some, unable to cope with larger ships, have fallen into disuse or sought alternative forms of commercial trade, others have had to resort to dredging. Rotterdam, the world's largest and most

Above Sewage works' settling pools may be unattractive, but they offer a perfect home to inland breeding black-necked stilts.

Right Pectoral sandpipers are American shorebirds that frequently cross the Atlantic to appear on European marshes. They are just as likely to turn up at a sewage plant as at a coastal marsh.

contaminants. Not every sewage plant uses the technology that requires (or allows, from a birder's perspective) open lagoons, but most have to store their treated sludge somewhere – and if it is stored in open ponds, as is usually the case, then it can provide another food source for birds. In addition to shorebirds, the more mundane scavengers are in attendance – starlings, grackles, house sparrows, crows and gulls, depending on the surrounding habitat.

As the sewage ponds show, human activity is not always bad for birds. Near the ocean and large lakes, shopping mall parking lots are favored nighttime roosts for gulls (and if there are fast-food restaurants in the mall, the pickings for spilled French fries can be good, too). Sod farms, where high-quality grass is grown for lawns, cemeteries and golf courses, provide good birding during migration for certain plovers, especially lesser golden and black-bellied. One can hardly conceive of a more barren habitat (save only for asphalt),

successful port, is a prime example.

Dredging deep water channels at the mouths of rivers is not only a continuous and expensive process, it also produces thousands of tons of unwanted silt. The usual method of getting rid of this is to pump it into lagoons alongside the estuary, creating wonderful brackish marshes which are absolutely perfect for waders. Being adjacent to what are already excellent wader habitats, these lagoons act both as feeding grounds and as roosts for the birds.

OASES IN CONCRETE

periods is remarkably rewarding no matter where it is, but when the area is confined by the sea, or a concrete sea, one has a feeling that one has probably seen all the birds that are present. In this way a daily visit can show just how the park is being used by migrant birds. Some days may be rather thin, but then there will be a small rush followed by a tail off until the next arrival.

At one time there were several of us in London each watching his own park throughout the spring and autumn. Each kept a daily tally of species and numbers and it was my job to draw all of these together and see what correlations could be made. Numbers, even of the commoner warblers, were

Left Stanley Park in Vancouver, British Columbia is an oasis not only to its human population, but also to small birds migrating along the Pacific coast.

Below Yellow warblers are among many birds that fall out in urban areas, where they take refuge in parks and gardens.

"An island in a sea of concrete" is a pretty apt description of a bird's eye view of a downtown park. Migrating birds finding themselves over a city which might measure as much as 20 miles across, are faced with a sea of inhospitable concrete broken here and there by islands of greenery. And just as they will seek out land rather than alight on the sea, so they are drawn to parks rather than making a landing on buildings or streets. In New York, Central Park has long been known as a splendid area for migrating warblers, flycatchers, vireos and thrushes, particularly in spring. Many city-based birders make an early morning excursion an integral part of their working day and "The Ramble" in the Park often holds almost as many birders as joggers.

Working a confined area during migration

generally small, but it was surprising how the occurrence of high figures in one park coincided with similar figures in the others. Even the different species coincided quite well, showing that birds were flying over the city during quite specific periods. The nearest proper bird observatory was at Dungeness on the Kent coast, and the warden kindly sent me the daily tally for comparison. To my surprise even this fitted the pattern, with peaks and valleys of the common migrants fitting nicely together. To my chagrin there were, of course, records of rarities scattered through the season at this prime coastal site, none of which had come my way in London. Many were birds that I could easily have gone to see had I not been so preoccupied with my own local patch. But, for a while, the daily changes in the number of relatively common birds provided far more satisfaction than the sight of the odd rarity or two. Now I am fortunate enough to live in the country and the daily round takes me no further than a brief tour of the garden, farm buildings and hedgerows. Without being in any way special, this small area of southern England produces all the common migrants in varying numbers day by day.

Some parks may be too big for a thorough daily exploration, but a good walk around will show which areas are best for migrants. Try the pond for a start, especially if it has natural banks and a bushy island. One of my earliest attempts to explore the "Fair Isle in London" concept saw me watching just such a bushy island in a small pond at the same time as the park keepers were making a routine visit for some reason best known to themselves. Landing at one end of the apparently

Left From the air New York seems to offer nothing to any migrant bird. Yet even in this apparently inhospitable environment the occasional bird, like this hermit thrush (**above**), may find a city tree, where it can rest and recover, even if the tree is planted in a pot.

bird-free island they acted as unintentional beaters producing a string of different warblers at the other end as I watched. Finally they flushed out a gorgeous spring male pied flycatcher. I never missed the pond again.

Bushy areas, particularly where avenues of low trees come together in a "cross-roads" are often more productive than solitary trees. The best section of all in my park was a "natural" area of birches and hawthorn creating a clump no more than 100 by 50 yards. Yet morning after morning this tiny patch of cover produced most of the

regular summer visitors to southern England and others besides. Willow warblers and chiffchaffs, whitethroats and lesser whitethroats, redstarts and spotted flycatchers all put in their appearances. Once even a wood warbler, always a scarce migrant, rattled away beneath the canopy. Sadly, this was always an awkward area to work - the ladies' and gentlemens' toilets were sited right in the middle.

If there is one piece of advice that can be offered to the potential park birder it is be persistent; please don't judge a park on a single day, at the wrong time of the year and, in particular, don't compare it with a coastal hot spot.

FRESH WATER: MOUNTAIN STREAMS

In the Rocky Mountains, a mountain stream may pass through many life zones, the stratified communities of plants and animals that exist only at specific altitudes. Thus, the same stream may play host to a flock of bathing rosy finches in high, alpine areas, and varied thrushes several thousand feet lower, in the thick coniferous forests. There are some western birds, though, that are closely linked to streams regardless of the surrounding habitat. Of them, the American dipper is one of the most intriguing. Small and nondescript, with a

Above The dipper is a specialist of tumbling streams where it swims and wades with consummate ease. Its wedge-like shape is an aid to submersion.

Left The spotted sandpiper spends its summers along the margins of fast flowing streams, where its continuous bobbing seems to merge with the bubbling water.

stubby tail and short beak, the dipper is plain gray – a bird for which most people would not spare a second glance. It is found along the verges of streams, bobbing its rear energetically as its flits from rock to rock, probing for insects.

The foraging dipper

Suddenly, the dipper will plunge headfirst into the foaming stream. This is not usual behavior for a songbird – but it is for the dipper, which is as much at home foraging on the bottom of a raging mountain stream as it is hunting along its banks. Adaptation has helped it cope with an aquatic lifestyle. Its plumage is very dense and quite waterproof, aided by large glands that provide oil for preening. A set of scales over the nostrils

closes off the breathing passages when the dipper is submerged. Its short wings make it a poor flier in the air, but allow the bird to "fly" underwater – a trait shared by other small-winged waterbirds like puffins and auks. As might be expected, the dipper even nests close to water, building a dome of moss with an entrance hole on the side, usually on a cliff face just above the water, or behind a waterfall.

Dippers are common across much of the West, but another fast-water specialist is not. The harlequin duck is one of the most bizarrely colored species of waterfowl in North America; drakes are blue-gray, with patches of black-edged white on the face, neck, and sides, and sporting chestnut-red flanks. It is found from the Rockies north through Canada and Alaska, and in localized areas

across subarctic Canada to Labrador. Harlequins prefer fast, turbulent waters that flow through thick conifer forests, nesting on the ground within sight of the river. They move with unmatched skill in the dangerously fast waters, disappearing beneath the roil only to pop up again far downstream, having gleaned crustaceans and insects from the bottom. Because they are larger than the dipper (thus requiring more food), harlequin ducks are usually found on rivers that drain lakes, and so support more invertebrate life than rivers without lakes. The LeHardy Rapids of the Yellowstone River in Yellowstone National Park, and McDonald Creek in Glaciers National Park, Montana, are two of the more accessible places where harlequin ducks may be found.

Tail "bobbers"

For reasons that have never been explained, many of the birds found along the water's edge bob their tails up and down, like the dipper. Perhaps the most energetic "bobber" is the spotted sandpiper, closely related to the common sandpiper of Eurasia. A shade over seven inches long, the

Above Harlequin ducks are white water specialists that inhabit large rivers where they cascade among boulders. Their ability to survive such torrents creates a special niche that they have totally to themselves among northern duck. This male and female were photographed in Oregon.

spotted sandpiper has a relatively short neck and legs, giving it a more squat appearance than most shorebirds. Tail cocked high, it bobs and teeters as it forages in shallow water along the stream bank, its body in almost constant motion. Through the spring and summer breeding season, both sexes wear a scattering of black spots on their white breast.

Two other birds that teeter are the northern and Louisiana waterthrushes, actually members of the wood warbler family. Similar except for Louisiana's buffier belly, the two are rigidly segregated by habitat. The northern waterthrush prefers to nest in swamps, among the roots of trees that overhang standing water. The Louisiana (which, incidentally, ranges as far north as Maine and Ontario) also likes to build its nest among roots, but it insists on running water nearby.

LOWLAND RIVERS

As a river drops from the mountains to the lowlands, its speed diminishes. The white water rapids of the upper reaches are replaced by languid oxbows, the temperature warms and the rocky riverbed increasingly becomes one of mud and sand that can support plant life.

In this environment, so much less hostile than the raging torrent of the mountain river, a more diverse mix of plants and animals can make their home. Much of this life is hidden beneath the moving water – dragonfly larvae stalking smaller insects, schools of minnows darting in the shallows, bluegills, smallmouth bass and channel catfish in deeper water.

The belted king fisher

One of the signature tunes of the lowland river is the rattled call of the belted kingfisher, a common waterbird over virtually all of North America. A chunky bird with small legs, a shaggy crest, and an outsized bill, the kingfisher is unusual in a number of respects. It is one of the few North American birds to exhibit what is known as reverse sexual dimorphism – that is, the female is the more brightly colored of the sexes. Both male and female are blue-gray above and white below, with a broad gray band across the breast, but the female is marked with a second wide band of chestnut that crosses the belly. The kingfisher is also odd in its nesting habits, excavating a long tunnel in a streamside bank that ends in a cul-de-sac where

the chicks are raised on a constant diet of small fish. On the hunt, the kingfisher may perch on an overhanging branch or hover on the wing, but when prey is sighted, it plunges headfirst into the water, snatching its quarry with accuracy.

The habitat and its wildlife

As on the mountain stream, a river's birdlife has much to do with the surrounding habitat. In open country, where the river may be bounded with lush vegetation and marshes, common moorhens, king rails, tree swallows, and common yellowthroats will be found; along wooded riverbanks, the players change to northern orioles, warbling vireos, prothonotary warblers, and (in northern areas) winter wrens. Where mature trees female shares her mate's shape, including the distinctive crest, but not his brilliant hue. Beset by overshooting in the late 1800s, and by a lack of nest trees, the wood duck's numbers dropped perilously low in the early decades of this century. Protected by law, it began a slow recovery, but its fortunes improved dramatically when sportsmen and birdlovers started building and erecting wooden nest boxes that approximate natural cavities. Indeed, the fakes often surpass the real thing; a box fitted with predator shields provides much greater security for the nest. The wood duck, which, in common with waterfowl, has a tremendous reproductive capability, has bounced back to the extent that it is now one of the most common waterbirds in the Northeast, despite a closely regulated fall hunting season.

While not as rich in waterbirds as marshes and lakes, rivers provide homes for several other notable species. In northern regions the common merganser nests along wooded rivers, seeking – like the wood duck – hollow trees; the same habitat may attract nesting common goldeneyes and buffleheads. Great blue herons, green-backed herons, spotted sandpipers, and herring gulls are common over much of the continent.

Above left A belted kingfisher holds a fish crosswise in its bill, but will have to turn it headfirst before swallowing. Strangely, unlike southern countries, both North America and Europe have only one species each.

Above Goosanders are essentially river birds during the summer, building their nests in cavities nearby. Sadly, they feed on young trout and salmon and are widely regarded as enemies by fishermen.

provide nest holes, and oaks drop their acorns into the water, lives the wood duck, probably the most beautiful of North America's native waterfowl. The drake is an arresting mix of colors – glossy green head with white chin marks and a red eye, lemon-colored flanks, blue and maroon wing feathers, and a wine-colored breast. The

Above Moorhens prefer the slower, more mature rivers, where a good growth of emergent vegetation acts as a hiding place for their bulky nests.

PONDS AND LAKES

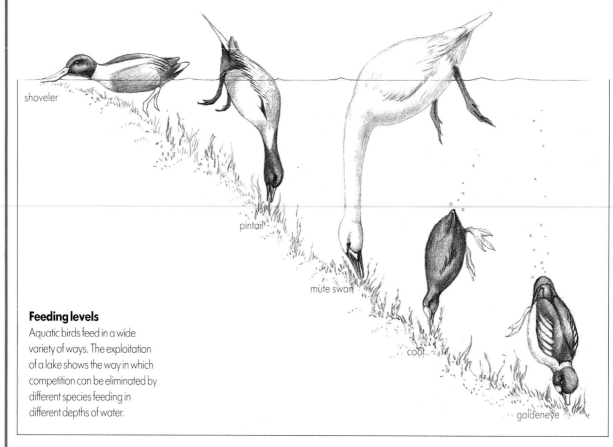

shoveler

pintail

mute swan

coot

goldeneye

Feeding levels
Aquatic birds feed in a wide variety of ways. The exploitation of a lake shows the way in which competition can be eliminated by different species feeding in different depths of water.

The essence of the northern lake country is summed up in the call of the common loon – a haunting distillation of the spirit of the deep, spruce-rimmed lakes of Canada and the northern United States.

Obviously, a cold, glacial lake in Quebec will have radically different birdlife from a shallow, warm pond in the South, or small, spring-fed lake in the arid West. Northern lakes tend to be deep, cold, and surrounded by forest, with little of the shallow marshes that attract so many waterbirds. The common loon thrives in such a habitat, though, diving deep for small fish and nesting in the small areas of emergent vegetation found along the shore.

An ancient race

The loons are an ancient race, among the most primitive of birds (in fact, they greatly resemble Hesperornis, an apparent link between reptiles and birds that lived some 70 million years ago). Life on and beneath the water has changed the loon. Its legs are placed far back on its body, providing excellent leverage for swimming but preventing it from walking on land; out of the water, a loon can only push itself along on its belly. Its eyes are capable of clear vision in both air and water, and because it has solid (rather than the normal hollow) bones, it has a specific gravity close to that of water, allowing it to force air out of its dense plumage and sink without commotion. Loons have been recorded diving more than 200 feet deep, but most of their hunting is in fairly shallow water where fish are most plentiful.

The loons carefully choose the nest site, usually along the shore of a small island where predators like raccoons and mink are less likely to be found. Both sexes build the platform of dead reeds and vegetation, constructing it right at the water's edges so that they have only to push themselves out of the water and onto the eggs. The male and female share incubation duties for nearly a month before the two or three eggs hatch; the chicks leave the nest almost immediately, following their parents – and frequently hitching a ride on their backs, nestled between their folded wings. Unfortunately, common loon populations have been dropping drastically in recent years. The reasons are many, and complex. Acid rain has rendered many lakes almost lifeless, and the chicks – which feed on aquatic insects and small fish – cannot find enough to eat. In many areas, increasing human recreation interferes with the nesting season. Boaters innocently scare the adults from the nests, separate them from their chicks, or swamp the low-lying nest with powerboat wakes. Fortunately, conservationists are making progress, especially in New England and the Great Lake states, by educating lake users and providing floating, artificial nest islands that are less susceptible to swamping.

Bald eagles and ospreys

The loons share their northern home with two other birds that prefer the undisturbed wilderness: bald eagles and ospreys, which nest in the surrounding forests and feed on the walleye, perch, and bass of the lakes. Both species suffered badly in the 1960s from pesticide contamination (picked up on their southern wintering grounds), but tightened controls on pesticide use have brought a resurgence in the numbers of these magnificent raptors.

Ducks, geese and grebes

Lakes and ponds are critically valuable to many birds. Waterfowl – ducks, geese, and swans – are reliant on them for both nesting sites and feeding areas. By far the two most common and widespread species are the Canada goose and the mallard duck, which have proven adaptable to almost any wetland situation, from a wilderness lake to an urban park pond. Mallards have always been common, but the Canada goose population has experienced a phenomenal explosion, particularly in the East. Once restricted as a breeding species to the northern half of the continent, this commanding bird now nests as far south as the Carolinas – the result of transplanting wing-clipped flocks, which eventually form non-

migratory populations, and the presence of so much waste grain left in farm fields after the harvest. In many areas Canada geese are now year-round residents instead of winter visitors, and it often seems that every farm pond and lake has a few nesting pairs, which may only fly south for a month or two when the water is completely frozen. By most counts there are 11 subspecies and geographic races, from the giant Canada goose of the Midwest (once thought to be extinct but now in no danger) to the tiny cackling and Richardson's Canada's of the Arctic, barely the size of mallards.

In the middle latitudes, many of the larger lakes are slow to freeze in winter, and some stay open throughout the season. These act as magnets for wintering diving ducks, which can penetrate the deep water for food that lies beyond the range of dabbler ducks like mallards, which only tip head-down to feed. Lakes such as New York State's Finger Lakes may support occasionally huge flocks (called rafts) of redheads, scaup, ring-necked

Above Virtually any area of still freshwater will enable a mallard to rear its young, whether it be a pond, or a lake. This adaptability means that the mallard is one of the most common water fowl.

ducks, canvasbacks, common goldeneyes, and buffleheads; even though the birds may congregate in a few food-rich areas, the species rarely mingle, but rather gather by species into smaller flocks.

In the West, one of the attractions that lakes and ponds hold for birders is the five species of grebes that breed among the reed beds on their banks. The largest and most spectacular is the western grebe, nearly two feet long, with a ruby-red eye and a rapier-sharp yellow beak; it is found on large lakes, where pairs race across the water in their stunning courtship dance. The smaller eared grebe is more widely distributed, from the Canadian prairie provinces south to the Mexican border. In breeding plumage, the eared is dark, with golden "ear" plumes behind each eye. Across most of western Canada and Alaska, the eared grebe is replaced by the horned grebe, also about a foot long, with a reddish neck, dark head, and more compact, orange head tufts. The red-necked grebe is almost as big as the western, and is found from the Great Lakes northwest across Canada to Alaska; it sports a reddish chest and neck, dark head, and large, white facial patches. The last of the common grebes is the pied-billed grebe, a drab, duck-like bird that ranges over most of North America, wherever there are shallow, marshy ponds. Like all grebes it is an expert diver, slipping silently beneath the water to hunt for small fish and invertebrates.

From West to South

Western lakes, many of which are shallow and rich in nutrients, support a wide variety of birds. Double-crested cormorants, white pelicans, redheads, pintails, shovelers, cinnamon teal, and blue-winged ply the waters. Overhead may be a half-dozen species of swallows, including violet-green, cliff and tree swallows, catching damselflies and other emerging insects; joining them in the chase will be Franklin's gulls, while California gulls and Forster's terns watch for fish.

In the South, naturally, the cast list changes. Here the climate is kinder, the waters full of fish and other food. This is the stronghold of the wading birds, the herons and egrets. They range in size from the great blue heron, which stands four feet tall and has a wingspan of six feet, to the

Left Common loons frequent large lakes in the conifer zone of northern North America. Their wailing cries are a distinctive sound of these wild forests.

Above Northern boreal lakes are summer home to the horned grebe. Called the Slavonian grebe in Europe, this attractive bird is named after the golden horns of its summer plumage.

Right Large lowland lakes with a good growth of marginal vegetation meet the needs of the green-backed heron perfectly. They seldom perch openly, however.

diminutive green-backed heron, which stands less than a foot tall. In between are the delicate snowy egret, with its white plumage, black legs, and striking yellow feet; the black-crowned and yellow-crowned night-herons, both crepuscular hunters of the dawn and dusk; the little blue heron, which undergoes a complete plumage change from pure white juvenile to purplish-blue adult; the cattle egret, an Old World species that was blown to South America in the 1800s and has since spread across the hemisphere; and the great egret, white and almost as big as a great blue heron.

Two of the most unusual southern species are the fulvous and black-bellied whistling-ducks, goose-like birds with long legs and long necks. The fulvous whistling-duck is found along the Gulf coast and parts of Florida, while the black-bellied whistling-duck inhabits wooded ponds and streams in south Texas and (uncommonly) in southern Arizona. The fulvous whistling-duck is one of the most widely distributed waterfowl species in the world, found in North and South America, Africa, and India. Its nest is usually built among thick rushes on the marshy verge of ponds, while the black-bellied variety is a cavity nester, like the wood duck.

BOGS

A bog is a unique ecosystem – the last gasp of a glacial lake, which over thousands of years has been reclaimed by the forest. It is mysterious, with quaking soil, stunted trees, beautiful wildflowers, and birds found nowhere else.

"Biological islands"

Most of North America's bogs are located in the North and East, across Alaska, Canada, and New England, then sporadically down the Appalachians. A result of the last ice age, mountain bogs are often "biological islands," where plants

Below Boggy margins and undergrowth among northern forests are the favored breeding haunt of the winter wren. In winter these tiny birds move southward into parks, yards and a wide variety of wooded habitats.

Right Bogs among extensive forests of spruce are the breeding grounds of the palm warbler. Like many other ground-dwelling birds it wags its tail almost constantly.

and animals typical of areas much farther north still exist. A bog is formed when a pond or lake – usually carved by a glacier – begins to lose the fight against the encroaching forest. The water silts up with the accumulating debris of the surrounding woods, and such wetland plants as sphagnum moss and cranberries probe further and further toward the center, forming a thick mat right over the remaining water. Eventually, the pond is reduced to a tiny hole of open water at the

bog's center, and over time even that remnant will be gone. The bog itself is a series of plant successional stages, from water to dry forest, with peat – the decayed remains of sphagnum – forming the underlayer. Even in wooded bogs, the soil may be so spongy and wet that a person must keep moving to avoid sinking.

Because bogs were formed during the retreat of the glaciers, when the climate was a good deal cooler and wetter than it is today, southerly bogs are ecologically different from the areas in which they are found; a bog in Pennsylvania's Pocono

plateau, for instance, will have plants typical of southern Canada, not the Mid-Atlantic region. The same goes for birds. Species usually considered northern, like the yellow-bellied and olive-sided flycatchers, the winter wren, Nashville warbler, and golden-crowned kinglet, are found breeding in the spruce-cloaked bogs high on the Appalachians.

In the north itself, bogs have an equally great importance for birds like the Lincoln's sparrow and other bog-nesters. But unlike most wetland areas, bogs are not themselves rich in aquatic life. The soil and water are highly acidic, and there are few insects that can tolerate the low pH. (This combination of acidic water and minimal dissolved oxygen makes bogs an excellent repository for clues to the past. Bacterial action is almost nonexistent, explaining why perfectly preserved human sacrifices from the Bronze Age have been unearthed in Europe. To the naturalist, this trait is

of more concern because bogs preserve plant pollen over the millennia, leaving a record of forest composition since the last ice age.)

The palm warbler and the fly catcher

The palm warbler is one of the northern species that prefers to nest in and around bogs. A little more than five inches long, the palm is best known among most U.S. birders as one of the first of the spring migrants, passing through several weeks before the main warbler flights in May.

Nondescript, the palm warbler has a yellowish belly and a rusty cap (the belly is paler in western birds). It nests in a band from the Canadian Maritimes to the Northwest Territories, staying in the trackless spruce forests. But not just anywhere in the forests – the palm warbler rarely nests away from spruce bogs, where it buries its grass nest down in the wet sphagnum moss at the bog's

Above The yellow-bellied flycatcher is a typical inhabitant of bogs and damp conifer woods. Although it has a distinctive call, it should be identified with caution.

periphery, laying four or five brown-blotched eggs.

The yellow-bellied flycatcher, a common northern bog bird, belongs to the confusing Empidonax genus of flycatchers, which frustrate even experienced birders with their almost identical appearance. The yellow-bellied is small and greenish-yellow (in fact, its belly is really no more yellow than the other Empids), with a prominent eye ring and two wing bars. Its lazy che-lek call is the best way of differentiating it in migration, but on the breeding grounds, it is, with the alder flycatcher, the only Empid commonly found in bogs. Like most bog-dwellers the yellow-bellied is a ground-nester, secreting its nest beneath the concealing roots of a spruce tree.

SWAMPS AND MARSHES

Wetlands – both fresh and salt – are among the most fertile and productive ecosystems on the face of the Earth. They provide the breeding or feeding grounds for a staggering variety of wildlife, and so they are also the richest hunting ground for bird-watchers.

The words "marsh" and "swamp" are used interchangeably by most people, although they denote two very different kinds of area. Generally speaking, an open, treeless wetland with abundant reeds and rushes is considered a marsh, whereas a swamp is wooded, either completely or partially. Swamps are most common in the Southeast, while marshes are the rule over much of the Midwest and West.

Prairie potholes

The richest inland marsh region in North America is probably the prairie pothole country, stretching from the upper Midwest across the Canadian prairie provinces. Here, the retreating glaciers carved hundreds of thousands of small ponds,

Below A trumpeter swan adopts a threatening attitude toward an intruding scientist. Once virtually extinct, these natives of North America have been the subject of several reintroduction schemes.

Right Largest of the herons, the great blue heron is a wetland bird that is most at home among fish-rich marshes, where it hunts with patience and stealth.

known as potholes in local parlance. Shallow and sometimes wet only in spring, the potholes are justly famous for the incredible number of birds they support. Long known as North America's "duck factory," the pothole region provides nesting habitat for the bulk of the continent's waterfowl – dabblers like mallards, pintails, blue- and green-winged teal. American wigeon and shovelers, and such diving ducks as canvasbacks, redheads, scaup, and ruddy ducks. Unfortunately, the potholes themselves are a critically endangered resource. Draining for agriculture has destroyed

untold numbers of these prairie jewels; in some areas, as many as 80 percent have been ruined. Those that are not drained are often useless to wildlife, because farmers have plowed right to the water's edge, eliminating the upland vegetation where many of the ducks actually build their nests. Where rims of such vegetation remain, the nesting ducks are so closely concentrated that they are easy prey for the red fox, raccoons, and other predators. Pesticides, barnyard runoff, and other pollutants have defiled still more potholes. Worse

Left Black terns spend the summer on shallow marshes, where they lay their eggs on a floating platform of vegetation. In full summer plumage they are unmistakable; in fall and winter they are easily confused with other marsh terns.

yet, potholes rely on snowmelt and spring rains, and in years of severe drought – the norm through most of the 1980s – many are simply desiccated depressions. Attitudes are changing, however, and U.S. potholes are now strictly protected under federal law.

The potholes and western marshes produce more than just ducks. Sandhill cranes are a common marsh and grassland bird in the upper Midwest and mountain states, where their trumpeting calls rock the morning air in spring and summer. Smaller than the endangered whooping crane, the sandhill is gray, with a crimson cap on adult birds. Sandhills are shy birds, wary of human contact, although they have no hesitation about mingling with herds of cattle. Each crane pair assiduously defends its nesting territory, to which it returns season after season, building a mounded nest of sticks and marsh vegetation that rises above the water's surface.

The black tern is another familiar marsh dweller, breeding from the Great Lakes region west to California, wherever there are large expanses of reeds and cattails. Graceful in the air, the black tern is the only all-dark species to be found regularly on the continent; its small size and notched tail are further field marks. The attractively blotched, buffy eggs are laid in a low nest of marsh vegetation, usually built on tiny islands. Black terns are known for their colonial tendencies and their strenuous defense of the nest; a predator that wanders into a black tern colony is mercilessly bombed, the attacks punctuated with the terns piercing alarm scream. Often to be found sharing the marsh with the black tern are the common and Forester's terns, two similar white species. The Forester's, with a black-tipped orange bill, is found as far south as Colorado and California, while the common tern, with a red bill, is more northerly. All three – black, Forester's, and common – may feed in the same marsh, but their food habits help avoid competition. The common eats up to 90 percent small fish, the Forester's eats a mix of fish, insects, and invertebrates, while the tiny black tern is largely insectivorous, snatching dragonflies and other bugs out of the air.

A spectacular rarity

The most spectacular bird of the western marshes is also one of the rarest: the trumpeter was shot for its meat and its down, so that by the turn of the century it was considered extinct, or nearly so. Then in the 1930s a small, nonmigratory flock was discovered in the remote Centennial Valley of southwestern Montana, on a series of spring-fed lakes that rarely freeze, even in the bitter Rocky Mountain winter. The handful of birds were protected with the creation of Red Rock Lakes National Wildlife Refuge, and their numbers have grown dramatically in the 50 years since then. It was discovered, in the meantime that a much larger population exists in Canada and Alaska.) The largest of North America's waterfowl, the trumpeter gets its name from the resonant honk heard on the breeding marshes. Each pair stakes out a large territory that includes quiet, undisturbed wetlands for nesting, long stretches of open water for takeoff and landing, and lots of food – aquatic plants for the adults, and invertebrates for the protein-hungry cygnets. Today, trumpeter swans are being reintroduced in parts of their former range, especially around the Great Lakes, thus helping to mitigate some of the damage done to this species.

Below Shallow marshes are the home of the beautifully marked male blue-winged teal. In the eclipse plumage of fall, the blue inner wing is the best means of distinguishing this bird from the green-winged teal.

The swamps of the South

Both in distance and atmosphere, southern swamps are far removed from the open marshes of the prairies. Here, tea-brown water reflects the forms of bald cypress trees, tupelo, and magnolias, while Spanish moss festoons the branches. At night, the swamps ring with the weird hoots of the barred owl, a common southern resident. Large and round-headed, the barred owl is one of the few North American owls with brown, rather than yellow, eyes; the effect is of an almost cuddly bird. At night (and frequently during the day), those huge eyes help the owl hunt for small mammals, birds, crustaceans, and amphibians. The barred owl's call is a series of rapid, yelping hoots, usually transcribed as Who cooks for you, Who cooks for you-alllll, the final note drawn out into a descending laugh. The result is a night sound guaranteed to raise the neck hairs of anyone new to the swamp.

As might be expected, wading birds are very much in evidence in the swamps. A wide variety of herons and egrets hunt here, as well as the white ibis, with its attractive red face and legs, and the wood stork, now an endangered species, with around 10,000 restricted to the Gulf Coast and

Carolinas. Unlike wading birds that require abundant rain to insure feeding grounds, the stork suffers if there are not periodic dry spells to shrink lakes and concentrate fish where they can be easily caught for the chicks. The drastic alterations in habitat and water flow through much of the Florida peninsula have been bad for other waders, and disastrous for the wood stork, which will simply not breed if water levels are too high.

One of the most unusual southern swamp birds is the limpkin, a relative of the rails. About two feet long, the limpkin has a long neck, bill, and legs, which stand it in good stead as it hunts the swamps for frogs, insects and, especially, snails; its beak even has an asymmetrical tip that helps the limpkin extract stubborn snails from their shells. Because it lives in deep swamps and is often active at night, much is still unknown about the limpkin's life history. Its call, a wailing krr-ee-ow, is another hair-raising swamp sound, and has given this odd bird many descriptive names over the years: screamer, mourning widow, and crying bird among them. Even its common name, limpkin, is accurate, because of the bird's halting, almost injured way of walking.

There is plenty of color in a swamp, belying the mental image of a dank, gloomy place. Much of the color comes from birds, especially the lovely prothonotary warbler, one of the commoner songbirds of the southern swamp. Golden yellow with a greenish back, the prothonotary (named for the papal notary whose color of office is yellow) nests in hollow trees that overhang the still water.

It is here that the warbler hunts as well, foraging for insects among the fallen logs that lie half-submerged. In a typical southern swamp like Georgia's famous Okefenokee, other common birds include red-shouldered hawks, king rails, chuck-will's-widows, pileated woodpeckers, yellow-throated warblers, and parula warblers.

Left Evening at an American marsh with Canada geese. Shallow waters like this are always under pressure both from shooting and agriculture.

Below The male red-winged blackbird sings to proclaim its ownership of a territory in a reedy marsh and spreads its wings to show its boldly colored patches.

SALT WATER: CLIFF COLONIES

What one notices first is the smell – a heavy, fishy stench when the wind shifts. A few more steps, and the drop-off comes into view, sheer cliffs plunging down to the sea hundreds of feet below. Creep carefully to the edge and peer down, and the air seems to be alive with birds, floating in the abyss like dust motes in a sunbeam.

A visit to a seabird colony has to rank as one of the greatest experiences a birder can have. Whether the setting is a foggy Aleutian island, the rugged headlands of the Pacific Northwest, or one of the precipices of Maritime Canada, the sights, sounds and, yes, smells, are absolutely unforgettable.

Auks and other cliff-nesters

The most diverse group of cliff-nesters are the auks, including the familiar puffins. In a very real sense, these are the "penguins" of the northern hemisphere; evolution has adapted them for life in cold water in much the same way penguins, which arose from different ancestral stock, evolved in the southern hemisphere. Both groups have the ability to dive deep and swim rapidly for fish, the biggest difference being that the auks have retained wings capable of flight, although these are only just big

Left The endearing puffin, here seen with a neat billful of fish, is a cliff nesting bird that excavates its own burrow in soft soil on cliff tops and landslides.

Above Bird Rock in Togiak National Wildlife Refuge, California is a wheeling mass of guillemots, kittiwakes, and puffins.

enough to keep them aloft. In fact the great auk was flightless, a factor that contributed to its extinction at the hands of sailors.

Happily, the other auks have fared better. On the eastern coast of Canada, Atlantic puffins, razorbills, common and thick-billed murres, and small black guillemots breed, while Pacific coast birders have an even richer selection – horned and tufted puffins, six species of auklets, five murrelets, the pigeon guillemot, and both murres. Most are cliff breeders, although the marbled murrelet breeds in old growth coastal forests, and the least auklet lays its eggs on bouldery beaches.

One of the most striking things about any bird cliff is the rigid stratification of the species, each gannets, which may breed in immense colonies.

Puffins

No other seabird enchants as much as the puffins. With their round bodies, stubby wings, and comical, multicolored bills, puffins have a clownish air about them. Atlantic puffins, like the horned and tufted, grow their fancy beak sheaths

Left Horizontal sedimentary rocks, eroded by wind and wave, create perfect ledges for common murres. With good feeding nearby, these birds may nest in thousands.

Below Razorbills prefer to nest among broken rocks where they can hide their egg in a cavity. Not for them the open ledges used by the more abundant murres.

always found at its own level. Take a typical cliff in Newfoundland; right above the surf level, nesting back in rocky crevices, will be found the black guillemots. Next, building their seaweed nests on small ledges, are the black-legged kittiwakes, a small gull species. Above the kittiwakes come the murres, where ledges are wide and level enough for them to lay their single eggs, and scattered among them will be found much smaller numbers of razorbills, nesting in crevices. Near the cliff top, where there is a layer of soil, Atlantic puffins will have burrowed underground, their entrances often just down-slope from the nest of herring and greater black-backed gulls and a few northern fulmars. The highest nesting seabird of all is invariably the Leach's storm-petrel, a swallow-sized bird that burrows, frequently by the hundreds of thousands, beneath the spruce forests and meadows that top the cliffs. A few cliffs, like some on the Gaspé Peninsula of Quebec, have flat, treeless tops that are perfectly suited to nesting

fresh for each breeding season; at the end of the summer colorful plates are shed, revealing a smaller, duller bill that the bird carries for the winter. The males dig the breeding tunnels, which extend back and down for about three feet, ending in a small chamber where the single egg is laid. The incubation and nestling period is long: about one-and-a-half months until the egg hatches, then another month or more until the chick leaves the burrow. Both parents incubate, and later shuttle fish such as sand launce to the youngster. Arriving at the nest, the parents will be seen to be carrying a dozen or more of the long, thin launce, neatly arranged with their heads gripped in the bird's beak. For years speculation raged over how a puffin could keep the first fish in place when it opened its beak for the next catch. The answer seems to be that when the puffin opens its beak to snatch a launce, the others are clamped against the roof of its mouth by means of its tongue.

DUNES AND ROCK BEACHES

For most people, the creaky call of gulls is inextricably linked with the soft rush of water on a sand beach, and the dull boom of waves breaking as they hit the shore. Naturally, there is more than this to the birdlife of the beach, but gulls are its most immediately obvious facet.

The gulls of North America

There are 21 species of gulls that regularly breed in North America, but many of them are restricted to the high Arctic, coming south only in winter. A few stay along the temperate coasts year-round, however. The herring gull is the most ubiquitous of these – found along the Atlantic coast from Georgia to Newfoundland, and inland across most of Canada. On the West Coast, its place is taken by the western gull (the same size but more of a fish-eater than the omnivorous herring), and further north by the pale glaucous-winged gull of British

Below A least tern arrives at its nest among the dunes. Increasing disturbance by beachgoers has driven this delicate little bird from many of its favorite haunts.

Columbia and Alaska. The largest and most powerful gull on the continent is the great black-backed, a fixture of New England and the Maritimes that has recently begun a rapid range expansion south, now reaching the Carolinas as a breeding species. By far the most predatory of eastern gulls, the great black-backed is a proficient nest-robber, and has had therefore a serious impact on the distribution of terns and common eiders in some areas.

If beachgoers were to vote, the most popular "sea gull" would undoubtedly be the laughing gull, found from coastal Canada down the Atlantic seaboard to Texas. Its black hood and red bill are as distinctive and classy, as its ringing "laugh" that echoes down the shore. Laughing gulls are smaller and more delicately built than herring gulls, lighter on the wing, seemingly more buoyant. A colonial bird, laughing gulls may congregate in groups of more than a thousand, making nest scrapes beneath tufts of dune grass and small shrubs. There is safety in numbers, because their eggs and chicks are sought by many predators – herring and black-backed gulls, raccoons, fox, stray dogs, and others. The fledglings lack their parents' black hood, which they will not attain until age three; after the breeding season the adults too lose it for winter.

Threatened by humans

Human use (and misuse) of beaches has threatened several birds. The piping plover is a small, sandy-colored shorebird of the Atlantic coast and some inland areas, where it nests on the beach above the high-tide line. But because of dune buggies, swimmers, free-running dogs, kids with kites - in other words, a typical day at the beach – the nests are rarely successful, and the piping plover is an endangered species. On government-owned beaches, plover nesting areas are now being fenced off from human intrusion, but the action may be too little, too late. The least tern, which also demands open beaches, was just recovering from plume hunting in the 19th century when it, too, was hit with the human tide on its breeding grounds. Like the piping plover, this robin-sized bird is in steep decline. The roseate tern, a threatened species in the U.S., is even less widespread than the least, nesting in colonies on sandy beaches and spoils banks.

Migrating spectacle

During the spring and fall migration, beaches are important feeding grounds for many species of shorebirds. Nowhere is this more dramatically

Left Roseate terns are among the rarest and most beautiful of all seabirds. Their white plumage, washed with peach on the breast in summer, picks them out from the other more abundant terns.

Below Though they nest in a wide variety of different landscapes, the largest colonies of herring gulls occur among dunes. In Holland such a colony provided the ideal subject for the ornithologist Niko Tinbergen to study gull behavior.

illustrated than along the shores of the Delaware Bay, around the full moon at the end of May. Just as millions of sandpipers, plovers, and other shorebirds are heading north on a nonstop flight from South America, further millions of horseshoe crabs are coming onto the beaches to lay their eggs. The shorebirds, exhausted and in need of abundant food, find it in untold quantities. The flocks of feeding birds are awe-inspiring; in places, the beach simply disappears beneath a shifting blanket of tens of thousands of red knots, semipalmated sandpipers, ruddy turnstones, and other species. Thousands of laughing gulls join the crowd, providing a pulse-quickening spectacle for watching birders. The shorebirds will feed voraciously for about two weeks, effectively doubling their weight, and preparing themselves for the next big leap north to their Arctic breeding grounds. Such way-stations as the Delaware Bay are of critical importance to migrating shorebirds, and biologists – who have identified a network of such concentration points in North and South America – are working to protect the habitat, and thus the birds.

Not all beaches are gentle sand. Many, especially in the north and west, are rocky, and the

birdlife on them is different from that on the sand beaches to the south. The black oystercatcher of the Pacific coast lays its two or three eggs in a small depression scratched out of the beach gravel, while in the winter, the surfbird spends its time rock-hopping among boulders and sea-spray.

BETWEEN THE TIDES

Above Purple sandpipers are typical rocky shoreline waders that, even in the breeding season, prefer this habitat. With their dark plumage they are easily overlooked amid the tide wrack.

The intertidal zone – that stretch of shoreline that is land or water, depending on the stage of the tide – is one of the most interesting parts of the coast. On southern beaches, it may only be a fairly narrow strip of wet sand, while in New England or the Pacific Northwest, it is a dramatic swath of rock and cliff, populated with hardy animals and plants that can absorb the punishment of frigid water and drying sun.

The intertidal zone
Birds are only one segment of the intertidal food chain. Like the birds of the sea cliffs that nest in segregated layers, so the invertebrates of the tidal area live in rigid order, known as zonation, which varies with the latitude. Take a jetty or piling along the gentle Virginia beach; above the highest water mark, but within the splash zone of the waves, will be found tiny sea roaches (arthropods rather than insects) and blue-green algae. Further down, where the high tide comes twice a day, are gray and ivory barnacles, which can survive the long periods when the tide is down. Still lower are anemones and sea squirts, which cannot tolerate as much exposure to air as the barnacles, and so must stay down where the tide retreats only briefly. Oysters cluster just above the low tide mark, where the exposure to air is of shortest duration, while an assortment of sponges,

hydroids, and seaweeds live just below them, in the world of constant water. Go several hundred miles to the north, to the coast of Maine, and the players are different. The spray and splash zones at the top of the rocky shore are dominated by lichens and periwinkle snails; successive layers below hold rock barnacles, blue mussels, sea urchins, starfish, and a half-dozen or more types of rockweed and kelp.

For a bird adapted to its shifting character, the intertidal zone provides good living. Along a beach, sanderlings skitter in unison, chasing retreating waves and dodging the next to rush up the sand. They are looking for aquatic invertebrates, exposed for an instant by the tumbling water of the retreating wave, and the sanderlings probe and peck with frenetic activity. Over the winter and during migration, rich tide zones may attract a wide variety of shorebirds, all

Right The extraordinary bill of the long-billed curlew is used to probe into deep mud, though it also spends time along rocky shores picking up tide-stranded crustaceans.

Below right A migrant group of sanderlings beats a hasty retreat before an incoming wave. These fast running little shorebirds feed on planktonic creatures stranded by each retreating wave.

of them seemingly competing for the same food resource. But are they really competing?

Researchers have found that shorebirds exhibit what is known as "resource partitioning" as they feed – that is, each species hunts in a unique way or place, thus avoiding direct competition with other species. The partitioning has a lot to do with bills. The long-billed curlew, with its extraordinarily curved beak, can probe very deep into the sand for mollusks and marine worms that are beyond the reach of other shorebirds. The American oystercatcher feeds primarily in the exposed beds of mussels, oysters, and other shellfish, using its knife-like bill to slit the powerful muscle that holds an oyster's shells together. True to their name, turnstones flip small rocks and shells to expose insects. Dunlin and red knots probe muddy soils (at different depths), while least sandpipers stay at the high-tide mark and dowitchers wade into shallow water to hunt. In winter, the purple sandpiper searches among large jetty rocks, often in association with turnstones.

Because it floods with each high tide, the intertidal zone provides no permanent home for birds, but it does provide food for several other varieties besides shorebirds. Gulls patrol it regularly, of course, watching for anything remotely edible that the sea may cast up. In winter, horned larks, savannah sparrows, and seaside sparrows may stray down from the dune grass to peck among the seawrack at the high-tide line. Tree swallows are a common sight in summer, swooping low for insects buffeted about by the off-shore breeze, and fish crows and boat-tailed grackles are regular scavengers.

Left Despite their name, oystercatchers feed mainly on mussels that they open with their large orange bills. They are generally found along rocky shorelines, but also frequent estuaries and sandy shores.

ESTUARIES

Right The abundant clapper rail is the only member of its family to have adapted to a marine lifestyle. It inhabits the salt marshes that line the estuaries of rivers.

Baking under a hot July sun, the air full of buzzing mosquitoes and biting marsh flies, a salt marsh may not seem the most inviting place. Thick, smelly mud sucks at your legs, and all around you, a sea of cordgrass spreads in unrelieved greenery. But set aside preconceptions, for the salt marsh and its attendant estuaries are brimming with life, and are one of the most rewarding places for a bird watcher to visit.

An estuary and marsh are another part of the intertidal zone, but one distinct enough to warrant separate treatment. Here, protected from the onslaught of ocean waves, yet fed twice a day by the ebb and rise of the tide, is a fabulously complex habitat, often referred to as the richest in the world.

Cordgrass – the backbone of the marsh

The backbone of the salt marsh is cordgrass, one of the few flowering plants that can tolerate life in a salt bath. There are several species, the most prevalent of which in the intertidal zone is saltmarsh cordgrass. It grows in lush profusion, densely packed stems burgeoning out of the tidal muck, and in turn helping to hold the marsh soil in

place with a lattice of roots and stems. The cordgrass takes the raw material of the marsh – the nutrient-rich mud that is replenished by the regular tides – and turns it into growing life. The nutrient cycle that the cordgrass taps into is a complex one; rivers such as the Susquehanna drain vast inland areas, collecting sediment and nutrients with each rainfall. They carry this burden slowly down to the sea, depositing it in estuaries when their currents are finally lost to the tides. Then the sea takes up the task of distributing the nutrients; conflicting currents caused by tides, differences in temperature and salinity move the nutrient load around, eventually bringing much of it to the marshes. Many of the coastal salt marshes are actually a blend of salt, brackish, and fresh: fresh at the inland edge, where there is usually a creek of sweet water entering the marsh; brackish in the middle, where creek and tide meet; and salt along the shoreward edge, where the creek channel contains the periodic flux of the tide.

Researchers who have studied the salt marsh say it is as fecund as prime Midwest cornfields, producing as much as 10 tons of organic matter per acre per year. The cordgrass is one of the foundation stones of this food web, both as a living plant and as decaying vegetation that further enriches the mud. Among the cordgrass, and in the warm, shallow waters that swirl around its roots, all manner of creatures come to feed and spawn. The estuary is the cradle of the ocean, for even pelagic fish come there to breed, filling the water in hosts that rival the flocks of birds in the air.

Waders of the marshes
Salt marshes are rich with wading birds like great egrets, tricolored herons, reddish egrets, little blue herons and snowy egrets. Shorebirds like knots

Above Waders, such as these knot, are driven from their estuarine feeding grounds by an incoming tide and fly to a regular high tide roost where they often pack tightly together.

Right Red-breasted mergansers come in with the tide to feed on the fish that are themselves most active at this time.

and turnstones show up during migration. Terns – Forster's, common, gull-billed, Caspian, elegant (in the West), and Sandwich – are a trademark sight as they fly on gracefully pointed wings. One tern, the black skimmer, nests on sand beaches but frequently hunts the tidal channels that wind through the salt marsh. Most terns catch fish by hovering and diving, but the skimmer has evolved a completely different system not shared by any

other North American bird. Its bill, rather than being modest as in most terns, is greatly elongated, with the lower mandible noticeably longer than the upper. The skimmer flies just above the channel's surface, opens its mouth, and drags the tip of the lower bill through the water, plowing a furrow that may extend for 200 yards; then it does a neat wing-over and retraces its path. The first pass creates a commotion that attracts small fish, and when the skimmer comes back a second time, its bill snaps shut the instant it bumps against a minnow. As an aid to detecting fish, the lower bill is rich with nerve endings, and because of the wear and tear of slicing through the water every day, the lower bill also grows about twice as fast as its upper counterpart.

The secretive rails
The thick cordgrass stands provide cover for the marsh's most secretive residents, the rails. More than one birder, scanning for new species, has seen a furtive movement of something brown in the marsh and passed it off, with a shudder, as a rice rat. Chances are it was a rail, for these small, brownish creatures can sometimes seem more rodent than bird. Sit quietly at dawn or dusk, and you'll probably be rewarded with a glimpse of a chubby bird with long legs and neck, and a thin, downcurved beak. It will move rapidly, in a jerky movement reminiscent of a windup toy. On the East Coast, it will most likely be a clapper rail, by far the commonest of the salt marsh rails – about 14 inches long, buffy with dark, vertical bars on its flanks, and a kek-kek-kek-kek call that sounds like

pebbles being clicked together. If the tide is high the rail will simply swim like a duck from one side of the channel to the other, while on ground it skitters in and out of the concealing cordgrass. Rails rarely fly unless hard-pressed, which is why so few birders have seen them.

While the clapper rail is the only common salt marsh breeder, other species like the similar Virginia rail, and the shorter-billed sora, winter on the salt marsh. By far the most secretive rail is also the smallest, the sparrow-sized black rail, found in scattered locations on the East and Gulf coasts, as well as on a few inland marshes. Almost never seen, it gives itself away by its midnight call, a sharp kik-kee-doo, kik-kee-doo that is maddeningly hard to pinpoint. Birders have waded

into the mosquito-filled nighttime marsh, had a black rail calling within two feet of them - and not seen it. Actually, such behavior is irresponsible, because there is a very real chance of trampling both the bird and its nest, and so it is far better to listen at a discreet distance.

The estuary in winter

Estuaries are as important in winter as they are in

Below Brent are the most estuarine of all the geese. They graze on beds of the seaweed zostera, but are also quite at home on nearby grasslands.

Right Black skimmers, seen here at their Texan nesting colony, are typically found at the mouths of large rivers where they practice their unique fishing technique.

feed on eelgrass, a flowering plant that is able to grow completely submerged in salt water. Unfortunately, eelgrass is prone to periodic die-offs, when disease decimates the beds. Brant, and other waterfowl that rely on eelgrass, are still recovering from a major dieback in the 1930s.

The list of waterfowl that winter on estuaries is a long one: Canada geese and tundra swans, American wigeon, canvasbacks, redheads, ring-necked ducks, lesser and greater scaup, buffleheads, common and Barrow's goldeneyes, ruddy ducks, red-breasted mergansers, and the beautiful hooded merganser. Looking like geese

Above Shorebirds frequently form mixed flocks on estuaries, especially when flighting between their feeding and roosting grounds. The long-billed birds are godwits, but there are also gray plovers with short bills.

Right Turnstones are shoreline birds that feed on estuaries, as well as rocky shores, throughout the year. This bird is in summer plumage at Churchill in Canada.

summer. Waterfowl in particular rely on them for their abundant food, shallow water, and ice-free conditions. Each fall, tens of thousands of snowgeese pour into New Jersey, Delaware, and Maryland, spending the winter feeding on cordgrass. In fact, some of the coastal wildlife refuges, like Brigantine in New Jersey, have experienced problems with burgeoning flocks denuding large areas of marsh, leaving them vulnerable to erosion without their protective layer of cordgrass. On the wider coves and bays, brant – a small dark goose found on both coasts –

but not related to them, double-crested cormorants fly in ragged Vs over the water, or dive deep in the muddy bays for fish. Out over the winter-brown cordgrass, northern harriers and short-eared owls watch for mice and rats, occasionally taking a sharp-tailed sparrow or a seaside sparrow that gets careless.

THE OPEN OCEAN

The open ocean hardly seems a likely place to find birds, but it is – which accounts for the growing popularity of pelagic birding trips, far offshore to the deep-water canyons of the continental shelf. Here, hardy birders willing to suffer seasickness and biting chill can expect to see birds that come to shore only briefly to breed – birds, indeed, that never touch foot to land in the Northern Hemisphere.

Pelagic birds

The albatrosses, shearwaters, and petrels are the most numerous of these pelagic birds. Fairly

Left Manx shearwaters come to land only to breed, and spend the rest of their year roaming the oceans, often well out of sight of land.

Above Old squaws are purely seaduck in winter, occurring in large rafts usually within sight of land, but coming closer inshore to feed on a rising tide.

primitive (at least when compared to songbirds), they are nevertheless finely tuned to their oceanic lifestyle. The albatross is unmatched for its grace in soaring, but not with the gentle circles of a hawk in a rising thermal of hot air. Instead, the albatross practices a method of flight known as dynamic soaring, gaining its lift by zigzagging across the rising spurts of air created as the seawind is deflected upward by waves. The albatross wings are so long and thin – between seven and 10 feet in most species – that they produce very little drag, thus increasing their soaring efficiency still more.

Solving problems

One of the hardest problems for a seabird to solve is that of water. The old adage "Water, water everywhere, but not a drop to drink," applies to seabirds as well as sailors – but while humans cannot drink salt water, seabirds can. The problem is the relative salt balance between ocean water and the animal's body fluids; because ocean water is three times saltier, a normal bird's kidneys would have to void two liters of water for every one liter of seawater drunk, leading to dehydration and death. But oceanic birds like petrels, puffins, and eider ducks have special glands above their eyes that filter salt from the blood stream. Even birds such as mallards, which ordinarily live in fresh water, develop enlarged salt glands if they spend any length of time drinking salt water.

The presence of salt glands and the ability to soar for hours has opened the ocean to birds. The largest are the albatrosses, of which only two, the black-footed and the Laysan, are seen in any numbers off the Pacific coast; yellow-nosed and black-browed albatrosses are rare off New England. Far more numerous are the shearwaters, named for their wave-skimming flight, and the petrels and tiny storm-petrels. At first glance, a shearwater can be mistaken for a gull with unusually long wings, but shearwaters are in a family by themselves, the Procellariidae, or tube-noses. Most are fish-eaters, snatching their prey from the surface or dipping just beneath the surface of the waves.

Shearwaters are totally at home on the open sea, hundreds of miles from land – land that they do not see for months on end. They are true world travelers, ranging thousands of miles from their breeding islands, treating the ocean as a single feeding ground. A pelagic birder in the North Atlantic is liable to spot greater shearwaters that breed on the islands off Argentina, Cory's shearwaters from the Mediterranean, and Manx shearwaters from the Azores or Iceland. The story is the same in the Pacific, with flesh-footed, short-tailed, and sooty shearwaters from Australia and New Zealand spending the austral winter feeding off the coast of California. The swallow-like storm-petrels follow much the same lifestyle, despite their small size; by some accounts, they are among the most numerous birds in the world. Not much larger are the red-necked and red phalaropes, shorebirds that breed on the arctic tundra but winter at sea, eating tiny crustaceans.

Sea duck

One group of rugged waterfowl, the sea ducks, take to the ocean in winter. While they are often within sight of land (and in the view of birders equipped with spotting 'scopes), they do not come ashore. Their numbers include common and king eiders, the harlequin duck, oldsquaw, and the three scoters – black, white-winged, and surf. Of the three, the surf scoter is certainly the most striking. The drake is velvety black, with a white neck patch and another on the forehead; the bill is large and garishly marked in orange, black, and white. It is hardly surprising that the New England fishermen called this bird "skunkhead."

Above Decidedly rare in Europe, surf scoters may gather in considerable flocks at favored locations in North America like these off Vancouver Island.

Right Wilson's petrels breed only in the southern hemisphere, but then make huge loop migrations into the North Atlantic.

CULTIVATION: FARMS

Change that is bad for one set of creatures is usually good for another. When the great forests east of the Mississippi were cleared in the 18th and 19th centuries, it was a disaster for woodland species like elk, mountain lions, and wolves. But for birds and mammals that prefer open land, it was a bonanza.

Farms may look natural, with their rolling fields of corn and wheat, but they are as much a product of human activities as any skyscraper. Just as some birds have learned to adjust to life in the city, so have many made themselves at home in farmland.

The ubiquitous swallow

A summer field doesn't seem complete without swallows scything through the air, making snap turns and daring loops as they hawk insects from the sky. Barn swallows are among the most widely known birds on the continent, in large part because they are found almost everywhere, and have no hesitation about living cheek-by-jowl with people. Virtually every farm building, from the newest tractor barn to the oldest shed, has a swallow nest or two tucked up in its rafters. Both sexes help build the cup, plastering mud and straw to the top of a beam, then lining it with grass and feathers to hold the four or five eggs. Barn swallows are quick to explore any opening in a building, and just as quick to start building a nest – something many garage owners have discovered after leaving the main door open during the day. The chicks grow fast on a steady diet of insects,

Left Barn swallows, as their name implies, are most abundant around farmsteads where they build their cup-shaped, mud nests inside barns and outhouses.

Top The kestrel is a familiar bird of town and countryside that is most numerous on open grass-lands where its small mammal and insect prey is abundant.

Above The eastern meadowlark is the typical grassland bird of much of southern and eastern North America.

and leave the nest in just a little more than two weeks on unsteady wings. Like their parents they have forked tails, but their color is much paler and duller than the adults'.

Other farm birds

Swallows are not the only common farm birds, however. A farm morning is alive with bird song: the Spring-of-the-year phrase of the eastern meadowlark, the caw of crows, the calls of bluebirds, robins, chipping sparrows, cardinals, grackles, and orioles. Even within the carefully tended confines of the farm, there are many habitats, many niches. The orioles stick to the trees – northern orioles up in the tall shade trees around the farmhouses, the maples and graceful elms, while the orchard orioles prefer fruit trees, groves and thickets along streams. House sparrows and starlings are common farm birds, but rarely stray far from the cluster of buildings where they can catch bugs and gather great wads of trash to make their nests. The overgrown patches at the ends of fields hold ring-necked pheasants in the fall, catbirds in the summer, and song sparrows all year.

Open hunting grounds

The open fields are the hunting ground of the American kestrel, the continent's smallest and most colorful falcon. The same size as a dove, this raptor is a hunter of grasshoppers and small rodents, spotted from a convenient telephone-wire perch or while hovering on furiously beating wings. It is also one of the few birds of prey that has differently colored sexes. The male has blue-gray wings where the female's are rusty, and her tail is barred with black where his is not. They are cavity nesters that will readily accept manmade boxes – not to mention a hole under the eaves of an abandoned barn.

The farm buildings are home for the barn owl as well; this is one of the most far-flung raptors in the world. Found on every continent except Antarctica, the barn owl is very much at home with people; while it will not hesitate to nest in a hollow tree, it is most often found raising its chicks in a silo or barn loft. The long-faced young, covered with down and dark, greasy-looking quills, quickly grow into pale beauties, cloaked in white-and-gold plumage, each with the barn owl's distinctive heart-shaped face. Far from mere decoration, the odd facial disk is vitally important to the owl, for the feathers that form it bounce sound waves into the bird's highly sensitive ears. The barn owl is the most nocturnal of the owls, and although its night vision is superb, it relies mostly on its hearing to capture voles, mice, and rats in the unmowed fields beyond the barn door.

The fencerows that stitch the fields together like a seam are pathways for many birds. Pheasants need them as routes from field to field, safe from the sight of foxes or hawks, and during the winter, when they break the force of windswept snow. Their tangles of chokecherry, greenbrier, poison ivy, and wild grapes are a feast for such fruit-eating birds as cedar waxwings and mockingbirds, and the same dense thickets and dead trees are prime nesting grounds for yellow warblers, blue grosbeaks, house wrens, and brown thrashers.

Left Barn owls are nocturnal predators around many farms, though they do require breeding sites that are safe from disturbance. For preference they also need overgrown or neglected grasslands over which to hunt.

GRASSLANDS

Above Short-eared owls are frequently found hunting during daylight over agricultural land. They are small mammal specialists and their flap and glide flight is similar to that of the northern harrier.

Perhaps more than any other ecosystem, the prairies of North America have undergone dramatic changes since they were first converted to agriculture in the 19th century. In particular, the tallgrass prairies of the eastern Midwest, where buffalo grass and big bluestem once grew so high and lush that they obscured men on horseback, have all but vanished beneath fields of wheat, corn, and soybeans.

An altered habitat

Such overwhelming alterations have obviously had a serious impact on the many species of birds that once nested in the grasslands. Some, like the greater prairie chicken, have been unable to adapt to the new agricultural conditions, and have disappeared across much of their former range. Others, like the eastern meadowlark, have taken the changes in their stride, and remain a common fixture of pastures and meadows.

Although the greatest areas of grassland lie in the prairie regions of the Midwest and West, there are grassland birds almost everywhere in North America. In the East, fallow fields attract meadowlarks, bobolinks, upland sandpipers, ring-necked pheasants and a number of sparrows – grasshopper, vesper. Henslow's, and savannah. A fairly recent addition to the list of grassland

birds is the red-winged blackbird, long considered strictly a wetlands nester. With a vast new breeding territory at its disposal, the blackbird responded with a dramatic rise in population.

Eastern hazards

Especially in the East, mowing for hay is one of the biggest hazards facing grassland species. Alfalfa and cool-weather pasture grasses are mowed in early summer, at the peak of the breeding season, and the mower's blades destroy nests, chicks, and incubating adults. Many birds attempt a second nesting after the cover has regrown, but frequently fall victim to exactly the same disaster when the farmer takes in a second crop later in the summer. Biologists point to early mowing as one of the prime reasons for the precipitous decline of the ring-necked pheasant in the East, and have also linked the mowing to drops in populations of other grass-nesting songbirds.

Prairie diversity

The East was once dominated by dense forest, and

open-country birds may be a fairly recent addition of the past two centuries. In the prairie regions, where the sea of grass has always held sway, the diversity of grassland birds is understandably greater. Western meadowlarks – almost identical to the eastern in appearance, but with a very different song – sing from fenceposts and telephone wires. Dickcissels sing their name from perches, or give their buzzy call notes in flight. This is a stronghold of the sparrows and finches, many of them adapted to nesting on the ground: Baird's sparrow in dry prairies of the northern Plains; Le Conte's sparrow in wet fields; lark sparrows, with their bold russet, black, and white face patterns; Cassin's sparrow, a grayish species of the southern Plains; chestnut-collared longspurs breeding where the grass is thick and tall, McCowan's longspurs in drier northern habitat. The unmistakable male lark bunting, coal-black save for large white wing-patches, is a common sight as he makes song-flights over much of the continent's midsection.

Species of many other bird families have made the grasslands their home. Killdeer, mountain plovers and upland sandpipers are three shorebirds that have made the transition to grassy environments, each with a distinct preference – killdeer in fields, pastures, and lawns, mountain plovers in high, dry plains east of the Rockies, and upland sandpipers in thicker grasslands. There are predators as well: short-eared owls and northern harriers both nest on the ground in grasslands, while Swainson's hawks and ferruginous hawks nest elsewhere but hunt the open lands. Gamebirds are well represented by greater and lesser prairie chickens (the latter a rare bird of the southern Plains), northern bobwhite in the East, and ring-necked pheasants across much of the northern U.S. and southern Canada.

In the winter, a fresh influx of northern species swirls into the open country – American tree sparrows, Lapland longspurs, snow buntings and, rarely, massive snowy owls down from the Arctic.

Above left Introduced from Europe for its sporting qualities, the ringnecked pheasant is widespread over agricultural land where its numbers are boosted annually by birds released for shooting.

Left The northern harrier is widespread from tundra to pastureland. It hunts low over the ground with alternate bouts of flapping and gliding and is an opportunist feeder.

BRUSHY AREAS

Plant succession is a predictable process. Clear land down to bare earth and leave it, and before too many days are past, a faint blush of green will show as weed seeds begin to sprout, or send up shoots from the damaged root stock. In a year or two, the ground will be awash in dandelions, ragweed and other "pioneer" species, the plants that are able to tolerate the dry, sunsoaked conditions. They, in turn, add humus to the soil, protecting it from the drying sun and keeping it cooler. A few years later, the first shrubs appear – usually mulberries, sumac, chokecherries, red-twigged dogwoods and other fruiting plants that birds enjoy. This is no accident, of course, because the seeds arrived in bird droppings in the first place. Provided there is no further disturbance, 10 or 15 years after the land was cleared there will be the beginnings of a thicket, and within 20, the plot will be a riot of saplings and shrubs struggling for a place in the sun. Allow the succession to continue, and the plot will revert to forest – but for now, let's freeze the action while everything is still

Left Named after its cat-like mewing calls, the gray catbird is a secretive bird of dense thickets and remains hidden for much of the time, making it difficult to spot.

Above Very similar to the closely related black-billed cuckoo, the yellow-billed cuckoo has distinctive rusty patches in the wing, and black and white outer tail feathers.

brush. The plot is too thick for open-land birds like grasshopper sparrows and meadowlarks, yet not mature enough for deep woods species like wood thrushes or red-eyed vireos.

Brush specialists
Seldom seen but easily heard, the yellow-billed cuckoo sulks thick in the brush, moving in a peculiarly disjointed way. The cuckoo is an elegant bird – creamy below, brown above, with a wash of chestnut on the wings and carrying a long tail edged in black and white. Yet it is the cuckoo's call, not its appearance, that draws attention to itself, a wooden cuc-cuc-cuc-kowlp-kwolp-kowlp so often heard before thunderstorms that its country name for years was "rain cow." Unlike the European cuckoo (of cuckoo-clock fame), the yellow-billed cuckoo is only rarely a parasite, laying its eggs in the nests of other birds. Usually it builds its own, a messy platform of twigs and leaves barely up to the task of containing the clutch. Yellow-billed cuckoos, with their close relative the black-billed cuckoo, are fond of hairy caterpillars like the gypsy moth and tent caterpillar, and will concentrate in a region suffering a major infestation of these insects. The next year, when the infestation is over, the cuckoos

Left Although a familiar garden bird over much of eastern and southern U.S., the colorful cardinal is originally an inhabitant of damp brush areas.

will be all but gone, following their gypsy lifestyle in search of more easy pickings.

Birders learned long ago that "pishing" – that is, making a sppsshh-sppsshh-sppsshh noise with the lips – often lures songbirds out of thick cover for a better look. It is almost a necessity for birding in brushy areas, where birds can be hard to see, although some species react to pishing better than others. One can pish all day at a willow flycatcher, a drab, brownish Empidonax, and not get a second glance from the bird. A catbird, on the other hand, will come boiling out of the shrubbery, tail cocked high, cussing a blue streak and ready for a fight. But the most enraged songbird of all is usually the male yellowthroat, a lovely warbler that is common over most of North America. Smaller than a sparrow, with a yellow breast, greenish back and black robber's mask, a pished yellowthroat will be

beside himself with anger – although why pishing should be so upsetting is anyone's guess. The warbler will often zoom up to a perch at roughly eye level, scattering sharp alarm notes as he comes. His normal posture is head-down, tail high, flicking his body from side to side with agitation. His mate, a plainer bird without the mask, will back him up, but from further down in the underbush, where their nest is hidden.

Birders who carefully comb brushy lands will find more than meets the eye. This is the place to look for northern bobwhite, whose clear calls are so beloved by many people, downy woodpeckers, chickadees and tufted titmice (if the saplings are on the tall side), Bewick's wrens, Carolina wrens, brown thrashers, white-eyed vireos, golden-winged warblers, chestnut-sided warblers and painted buntings (in the South).

FORESTS: DECIDUOUS

The wide temperate belt across North America, north of the subtropics and south of the boreal zone, is the domain of the deciduous woodlands. In the Northeast, the forest is a mix of oak, hickory, maple, ash and other broad-leaved species; along the waterways of the Plains, cottonwoods are dominant, while Western lowlands have oaks, ash, dogwoods, cottonwoods and aspen.

In spring, as the first hint of pink shows in the east, the wood thrush begins its morning song, a liquid series of flute-like notes that are famous for their beauty. The wood thrush is the most adaptable of the five forest thrushes, able to live in wooded suburban lots, mature forests and relatively brushy areas. The veery, a close relative, sticks to dense, wet deciduous woods. Beginning birders may have some trouble sorting out the two, but should quickly learn that the wood thrush has reddish-brown upperparts and a boldly spotted breast, while the veery is a bit grayer, with indistinct streaking on the buffy breast. The veery's song is also quite different – a series of descending *veer, veer, veer* notes.

Deciduous stratification
Stratification occurs in deciduous forests just as it does in the tropics, where some species are restricted to the canopy layer while others are limited to the ground. In the eastern deciduous woods, the wild turkey is the king of the forest

floor, a magnificent bird that has made a welcome recovery from near-extinction. An early casualty of forest clearing and colonial hunting, by the 19th century the turkey was decimated, but thanks to protection and intensive management, it is again common over most of its range. By late winter, the male gobblers are feeling the urge to begin mating, a stirring that hits full stride by April and May. Advertising his presence with strings of raucous gobbles, the male gathers a small harem of hens. To further his suit, the gobbler struts before the usually indifferent hens, fanning his tail, dragging his black-and-white wings and puffing out his bronze chest. Once mated, the drabber hens slip back to the forest to make a simple depression in

Left Deciduous woods are among the richest of all bird habitats. Although these young wood thrushes are heavily speckled they have the rufous upperparts that mark the adults.

Above The scarlet tanager is one of only a handful of tanagers that have spread northward from their tropical origins. It is a common summer visitor to eastern deciduous forests.

the leaves and to lay their large clutch of eggs.

Other ground-dwelling birds include ruffed grouse, hooded, worm-eating and Kentucky warblers, ovenbirds, rufous-sided towhees, dark-eyed juncos and whip-poor-wills, the last an increasingly uncommon relative of the nighthawk.

The treetop canopy
A good pair of ears are a prerequisite for finding the birds of the treetop canopy since they are

Left The large eyes of the eastern screech owl are an adaption to a strictly nocturnal life style. These birds inhabit a wide range of habitats, but are probably most at home among broad-leaved woods.

Below Wild turkeys, like this displaying male, were once abundant in woodland throughout the U.S. Although reduced in numbers and less widely distributed, they can still be found scratching a living from forest clearings.

usually heard before they are seen. The red-eyed vireo is known as the "preacher bird" because of its seemingly endless series of short, musical phrases. A birder with a keen ear can pick out the songs or calls of great-crested flycatchers, eastern wood-pewees, least and Acadian flycatchers, blue jays, yellow-throated and solitary vireos, a host of warblers such as blue-winged, black-and-white, black-throated blue, cerulean and yellow-throated, American redstart and rose-breasted grosbeaks.

The most stunning member, though, is certainly the male scarlet tanager. Fiery-red, with black wings and tail, a tanager simply glows when a shaft of sun highlights him against the dark green vegetation. In fall the male takes on the plain olive-yellow plumage of his mate for the long migration to South America.

CONIFEROUS

Left Spruce grouse, like this male, are birds of the great northern conifer forests. In territorial display much is made of the fan-shaped tail and the red crown combs.

Above The Cape May warbler is a summer visitor to black spruce forests across the boreal zone of Canada. Like so many other small birds of conifer woods, its song is thin and high pitched.

One of the last great wildernesses in North America, the boreal forest is a vast, largely untracked region of spruce and fir. This is the home of the goshawk and spruce grouse, the gray jay, white-throated sparrow and bizarre crossbills.

It is tough birding. The height of the summer breeding season in June is also the peak of the insect season, when clouds of blackflies and mosquitoes can literally drive a person crazy.

Blackflies are small gnats that breed in the clean northern waters in multitudes that can number in the millions per acre. Like mosquitoes, the females need the added nutrition of blood to allow egg production, and although they usually rely on moose and bears, human victims are eagerly accepted. The bite forms a red welt that itches maddeningly for days, and a birder not protected with long sleeves, plenty of bug repellent and even a head net will soon have hundreds of bites.

Birding in the forest

Further adding to the difficulty in birding the boreal forest is its dense nature. The spruce trees form an impenetrable wall of living branches and dead logs, with thickets of blueberry and Labrador tea bushes choking any clearing. The trees themselves are draped with Usnea lichen, or old man's beard, an air plant that grows in gray masses like Spanish moss. But the birds, hard as they are to see, are here. The northern coniferous forest is where more than 15 species of wood warblers come to breed, including the beautiful Cape May and magnolia warblers. Great horned,

Above This northern goshawk cares for its young in a nest which, typically, is placed on a high branch against the trunk of a tall conifer. This large bird is often the most powerful avian predator in many conifer forests.

Right This male white-winged crossbill displays the bold white, double wing-bar from which it is named. There is also white on the tertial wing feathers and a pinker tone than its more familiar relative, the red crossbill.

great gray, boreal owls and northern hawk-owls patrol the night, while sharp-shinned hawks, goshawks and merlins hunt by day. Both spruce grouse and ruffed grouse are common, although the ruffed is found most often where birch and aspen grow in small thickets. The quavering "Oh Sweet Canada, Canada" of the white-throated sparrow is heard almost everywhere; flocks of evening grosbeaks, pine siskins, red and white-winged crossbills, purple finches and pine grosbeaks are a regular sight, especially in fall. Winter is a bleak time of year, however, when the temperature drops far below zero and the forest is buried under snowdrifts. All but the hardiest birds have moved south, leaving the forest to the jays,

ravens, grouse, boreal chickadees, the black-backed and three-toed woodpeckers.

A change of scene

Far to the south is a completely different sort of conifer forest. The Coastal Plain, curving along the Gulf of Mexico and up the Atlantic seaboard from Texas through the Carolinas, is covered with forests of longleaf, shortleaf and loblolly pine that thrive in the dry, sandy soils. There is little similarity to the fauna of the boreal forest. Here, wild turkey and bobwhite, chuck-will's-widows and pine warblers are found, as is the endangered red-cockaded woodpecker, which nests only in pines infected with heart-rot. The zealous suppression of fire, coupled with the important Southern timber industry, have backed this species into a corner. The red-cockaded has the most interesting of woodpecker lifestyles; it lives in "clans," or family groups, made up of a pair, its young and a few unmated males, all of which defend a large territory against other red-cockaded woodpeckers. The red-cockaded shows a preference for longleaf pine, while the brown-headed nuthatch, another southern speciality, is found almost nowhere but in loblolly pine woods.

Not yet endangered, but declining over most of

its range, is the Bachman's sparrow, named for the Rev. John Bachman, a colleague of Audubon's who discovered this brown bird in the 1830s. Its other name, the pine-woods sparrow, gives an indication of its preferred habitat, although it is also found in oak woods. Timbering operations that opened large clearings prompted the Bachman's sparrow to expand its range to the northwest, up the Ohio River valley, but in recent years it has pulled back.

MIXED FORESTS

In biological terms, the "niche" an animal occupies is not a physical place (that is its habitat), but rather the way it lives within its habitat. The red-shouldered hawk, and the barred owl are hunters of small rodents, amphibians, reptiles, crayfish and birds. The only differences between their niches is that the hawk hunts by day, the owl by night.

Ecological analogs

These two birds are an example of what can be called ecological analogs – unrelated species that fill similar niches. There are many analogs between hawks and owls – red-tailed hawks and great horned owls, and American kestrels and eastern screech-owls, for example.

One mixed woods raptor that lacks an owl analog is the sharp-shinned hawk, the smallest and most agile of the accipiters, or forest hawks. Sharp-shinneds are a little larger than jays, with long tails and short, rounded wings for speed and maneuverability. They eat mostly birds, which are captured after frantic chases through the underbrush, with the hawk sometimes even pursuing its prey on foot through thick cover. Nothing strikes fear into woodland birds as much as the sudden appearance of a sharp-shinned; the songbirds cower among protective leaves, all the while giving piercing alarm calls to warn others that danger is among them.

Even the most jaded birder cannot help emitting a gasp of admiration when sighting the pileated woodpecker, North America's largest woodpecker (after the presumably extirpated ivory-billed) and an increasingly common bird in mature mixed and deciduous woodlands. As big and black as a crow, with a crimson crest and bold white markings, the pileated is a showstopper as it swoops through the forest or whacks away at a tree, sending wood chips in every direction. Its cuttings are distinctly rectangular, and are usually found in trees infested with large carpenter ants; the holes, in turn, provide cavities for squirrels, raccoons and owls.

The pileated has shown a remarkable resiliency;

Above This whip-poor-will is dozing away the day on its nest in a mixed wood in Michigan. Like the other nightjars, its camouflage is highly effective in protecting it from ground predators.

Right A female hairy woodpecker brings food to its nest hole. These are birds of mature woodland, often mixed, that are widespread in North America.

Feeding stations

Knowing which birds feed where
is often the key to locating them.
Even in a single tree different
species feed at different levels
and in different ways.

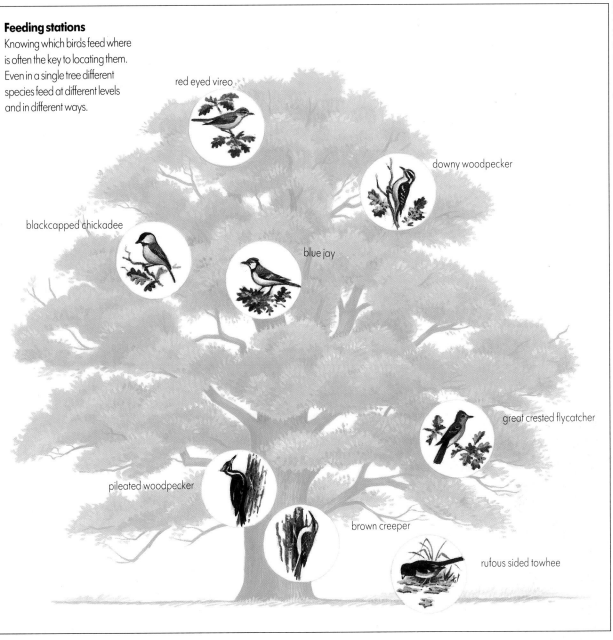

red eyed vireo

downy woodpecker

blackcapped chickadee

blue jay

great crested flycatcher

pileated woodpecker

brown creeper

rufous sided towhee

it was originally a bird of undisturbed, mature forests, which were the first to go under the ax when timbering operations hit full stride in the 19th century. In addition, the cock-o-the-woods, as it was called, was shot for food and for its red crest. By 1900 it was gone from almost all of the Northeast, but in the past 60 years it has adapted to smaller territories and less mature forests. That, coupled with the aging of the eastern woodlands (old trees for feeding and nesting), has allowed the pileated to make a strong rebound.

The mixed woods are home to many smaller birds as well. Parula, black-and-white, blackburnian and black-throated green warblers are fairly common mixed woods residents in the East, and in the South the summer tanager is most often found in oak-pine forests. Rose-red rather than the brilliant crimson of the scarlet tanager, the summer tanager also lacks the black wings and tail of its more northerly cousin.

SUBTROPICAL

This specialized habitat occurs in only a very few locations in the U.S. – on the extreme southern end of the Florida peninsula, and in south Texas near the mouth of the Rio Grande River.

The woodlands of south Florida and the Keys are of two types – drier upland forests of palms, mahoganies, lignumvitae, poisontree and gumbo-limbo trees, and tidal swamps of mangroves, and each type, obviously, has its own bird life. The entire region has been affected by its proximity to the islands of the Caribbean, and many tropical species have leapfrogged from island to island, reaching their northern limit here. One is the black-whiskered vireo, found in the Keys and along the Florida coast, as well as south through the West Indies. Looking like a red-eyed vireo with a

Below Found only in the Florida Everglades and Keys, the white-crowned pigeon nests among mangrove swamps and flights inland to feed on fruit trees.

Right The mangrove cuckoo is a typical inhabitant of the lush, mangrove swamps of coastal Florida. It is unknown elsewhere in the United States.

dark "mustache" mark, the black-whiskered vireo breeds in the mangrove swamps, building a typical vireo nest – a deep cup suspended between forks of a tree branch. Its appearance and call are so similar to a red-eyed vireo that it is likely to have been isolated for a fairly short time (evolutionarily speaking) and so has not diverged from its ancestral stock in many significant ways.

White-crowned pigeon

Another Caribbean species that excites northern birders is the white-crowned pigeon, a large, robust bird with blackish plumage and a shining, silvery cap. Unfortunately, the white-crowned pigeon has not had an easy time of it. Found only in the Caribbean basin, it has traditionally been pummeled by year-round shooting in many of its island homes, especially the Bahamas, where hunting is now greatly restricted. Migratory and erratic, white-crowneds move their breeding colonies from island to island if they are disturbed; a human intruding among the mangrove swamps where they nest will scare off the adults, leaving the eggs and chicks exposed to red-winged blackbirds, which kill them. In Florida, where the pigeon has been protected for decades, the problems are habitat loss (to dredging and filling of mangrove swamps) and a lack of food (the poisontree, whose fruits are avidly sought by pigeons, is detested by people because of its irritating sap).

Birds of the mangrove

The mangrove swamps where the pigeons nest are

strange, mysterious places. The mangrove is a unique tree that can grow in salt water, rising above the tideline on arching roots, which allow it to breathe, forming a dry-land community over shallow water. Roseate spoonbills, possibly the loveliest of Florida's birds, nest in the mangroves, filling the trees with their pink forms. Mangrove cuckoos, similar to the yellow-billed cuckoo but with black masks, are so adept at hiding in the swamps that virtually nothing is known about their breeding biology and life history. Other interesting birds of south Florida and the Keys include smooth-billed anis and gray kingbirds.

Rio Grande Valley

Along the Texas coast, the subtropical zone re-enters the U.S. in the lower Rio Grande Valley, just north of the Mexican border, although it is not all forest here. This region, too, has its avian exclusives, actually many more than the Keys

because this is mainland, rather than isolated islands. It is a long list – plain chachalacas, white-fronted doves, red-bellied pigeons, greater kiskadee flycatchers, the beautiful green jay, with its jade and turquoise plumage, the groove-billed ani (found elsewhere in Texas also) and the small olive sparrow. Couch's kingbird, once considered a subspecies of the tropical kingbird, inhabits woodlands along waterways. The Altamira oriole, a resplendent orange bird with a black face and throat, is increasingly common in protected forests of the lower Rio Grande. This is the only place in the U.S. where three species of kingfishers can be found – the widespread belted kingfisher, the small green kingfisher and the ringed kingfisher, which looks like a belted with a rusty belly, rather than just a belt of that color.

Below Dense subtropical forests offer breeding sites to the boldly-colored roseate spoonbill.

Not surprisingly this exotic bird is confined to the southernmost United States.

OPEN COUNTRY: SEMI-DESERT SCRUBLAND

Vast areas of the western U.S. are covered with arid scrublands – areas of little rainfall that are still too moist to qualify as true deserts. While there is some overlap with birds of the true desert and birds of better-watered areas, the scrublands have many species all their own.

The Harris' hawk, a most elegant raptor, lives in semi-arid brushland in Arizona, New Mexico and Texas. A large buteo, the adult Harris' is chocolate and reddish brown, with a white black-tipped tail. Long-legged and powerful, it hunts for rabbits at dawn and dusk, chasing them from cover and then snagging them on the run. There is nothing unusual in this approach, which many buteos use – but the Harris' hawk hunts in teams, a discovery made by Dr. James C. Bednarz, a raptor biologist working in New Mexico. Bednarz found that social

units consisting of a breeding pair, unmated adults and the immatures from the previous nesting season – often five or six birds – would surround vegetation that held desert cottontails and jackrabbits. While one hawk chases the rabbit from its refuge, the others gang up on the prey when it bolts. Then the entire group shares the meal – something almost unheard of among other raptors, which jealously guard their kills from all but their mates. Harris's hawks are unusual, too, in their breeding arrangements. The unmated adults may serve as "helpers" for a nesting pair, although many nests are tended only by the male and female. Researchers have found those with helpers do tend to have second broods more frequently, and raise somewhat bigger chicks.

Quail coveys

Head plumes bobbing as they walk, a covey of Gambel's quail are among the most endearing of the scrubland birds. Common over much of the Southwest and southern California, Gambel's quail are one of three confusingly similar western species; the California quail prefers woodier habitats and has a scaled pattern below, that the Gambel's lacks, while the mountain quail of the Pacific coast and the Northwest has long, straight head plumes that do not curve forward. More so than any other gamebird, the Gambel's quail is at home in arid environments, found in true desert as well as scrub. Its body processes food so as to extract every possible bit of moisture, and its digestive system is so efficient that only the barest minimum is excreted. Still, the Gambel's likes to stay where a permanent source of water is available, such as a stock tank or a small stream, and in autumn, after the breeding season is finished, flocks of up to 50 will gather in the vicinity of water. Overlapping some of the Gambel's quail's range is that of the scaled quail, an attractive bird found in dry areas from Arizona and Texas up into Colorado. It sports a crest rather than head plumes, and its body feathers are edged in black, giving it a scaly appearance.

Below Gambel's quail is a common bird of dry arid scrub wherever there is a permanent nearby water supply. It frequently forms flocks of 30 or more birds.

Other scrubland birds

Other common scrubland birds include the lesser nighthawk; the rock wren, a large, grayish wren

Left Dry canyons and brush are the favored haunt of the curve-billed thrasher in the border states of the southwest. Despite its restricted American distribution, it is a common bird of these habitats.

Below The beautifully marked Harris' hawk is found in arid scrub only along the Mexican border states. It is resident in yucca and saguaro country, but is slowly declining in numbers.

found over most the West; Bendire's thrasher, restricted to the Great Basin region; the curve-billed thrasher; mountain plover; and Cassin's sparrow, which prefers arid grasslands with brush. Costa's hummingbird, also a common desert species, lives in the dry scrub of Arizona and California. Only 31/2 inches long, the male Costa's has an iridescent purple head and throat patch, the latter extending back in two flanges like a giant set of sideburns; the female, as with most hummingbirds, is simply colored with a green back and white underparts. The iridescence that makes a hummingbird glow is a function of feather structure, rather than pigment. The feather is covered with a layer of special cells that refract light, breaking it into its spectral components. Some of the wavelengths are intensified while others are nullified, and the light that is bounced to a viewer's eye is brilliant – if the angle is right. Should the hummer move, changing the relationship between the angle of the sun and the viewer, the iridescence may be quenched, and the feathers will appear black.

DESERT

For most people, their image of the desert has been shaped by cowboy movies and television – a vast wasteland of searing heat, towering cactus and howling coyotes. The heat is certainly here, as are the cactus, and the coyotes. But the desert is not a wasteland. It has an abundance of life, shaped to the desert's rhythms of heat, cold, aridity and moisture.

The night cycle

Little stirs during the day, when the sun beats mercilessly. But as twilight comes the air cools quickly, since there is little atmospheric moisture to form heat-trapping clouds. This is the signal for mammals, birds, insects, reptiles and arthropods to emerge. A scant six inches long, the elf owl is the smallest owl in the world. Having spent the day in a hollow tree or saguaro, it hunts for insects, mostly beetles, grasshoppers, crickets and moths, although it will also take scorpions, which it disarms by biting off the dangerous stinger. Its

Above This gila woodpecker has excavated a hole in a saguaro in southern Arizona. These birds are largely responsible for the holes in this cactus, which are often used later by a variety of other birds.

Right The cactus wren is the largest of the North American wrens. It also constructs one of the largest nests, an untidy domed structure built amid the spiky protection offered by the cholla cactus.

unexpectedly loud call, a series of *churp* notes, is a common night sound in its rather restricted range in Arizona, New Mexico and Texas. The elf owl lays its two or three eggs in the old nest holes of woodpeckers.

The roadrunner

The roadrunner is one of the best-known desert birds, although it is also found in better-watered, scrubbier habitat. A giant cuckoo, the roadrunner is nearly two feet long, and it has left the trees for the ground; its long legs give it great speed and maneuverability, although not as much as its cartoon counterpart enjoys. Like all cuckoos the roadrunner has zygodactylous feet – that is, two toes point forward and two back, forming a large X. A similar arrangement is also found in woodpeckers, owls and parrots, and may give them better stability in running, perching and hunting. The roadrunner is a terrific hunter, feeding on lizards, snakes, insects, small birds and rodents, along with some fruit. At the onset of the breeding season, the male displays for his intended mate by raising his head and drooping

Left This roadrunner has caught a blue-bellied lizard. Although quite capable of dealing with sizeable snakes, this bird's reputation as a snake-killer has suffered some exaggeration.

Above This tiny elf owl is most active at dawn and dusk, but may frequently be seen peering from its nest hole in a saguaro.

his wings and tail, bowing, fanning his tail and singing a pigeonlike coo. The nest is a substantial structure of sticks, which is crowded in short order with four or five clamorous chicks. Interestingly, the male handles most of the incubation through the chilly nights, while the female lowers her body temperature as a way of conserving energy.

Birding at dawn

Desert birding is exciting, especially in the hour or two after sunrise, when the temperature is still comfortable and when the birds are most active. Cactus wrens sing, a rapid-fire *cha-cha-cha-cha-cha*, near the large, round twig nests they have built in cholla cactus; while the female incubates their first clutch, the male may build a new nest nearby for the second brood. The dapper black-throated sparrow, with an inky bib and white facial stripes, is a common sight among rocky slides. White-winged doves feed on the red saguaro fruits, brown-crested flycatchers hawk insects, and verdins, black-tailed gnatcatchers and phainopeplas can be found – although not each in every desert. The Southwest has three major desert regions: the Great Basin in the intermountain area, the Sonoran in southern California and Arizona, and the Chihuahuan in west Texas and southern New Mexico, and each has bird species all its own.

SAGEBRUSH FLATS

From a distance in bright light, the ground looks as though it were buried beneath a silvery snow, one that flickers and ripples in the breeze. Closer, and the covering takes on a greenish tinge, resolving itself at last into an endless expanse of brush – sagebrush, to be precise.

This most quintessentially Western of plants (there are several species, with big sagebrush the most common) prefers the rocky, semi–arid soils that dominate much of the area between the Continental Divide and the grasslands of the Plains. It is a hardy, woody shrub that reaches three or four feet in height, and gathers around it a community of birds and mammals found nowhere else.

Above In full display the male sage grouse makes much use of his sharply pointed tail feathers and puffs up his breast to expose the white ruff.

Left A male sage grouse has surrounded itself with a harem of admiring hens. In display he inflates the air sacs on his neck to produce a loud popping sound.

Sage grouse and horned lark

Herds of pronghorn antelope are the most visible sagebrush residents, but to birders, the most exciting is the sage grouse, the second largest (after the turkey) of our native upland gamebirds. Weighing as much as six pounds, sage grouse males are impressive birds, with dark brown plumage, a yellow comb over each eye and a thick white ruff that extends down from the neck and across the breast. In early spring, starting in March, the males gather at traditional leks, or courtship sites – usually barren areas where visibility is greatest. There they perform a complex, strutting dance in which the males drop their wings, fan their spiky tails and inflate air sacs in their chests, causing the white bibs to expand. It was the courtship displays of the sage grouse and the sharp-tailed grouse that many Plains Indian dances imitate.

The sage grouse is totally dependent on sage flats for its survival, especially in the winter, when it eats almost nothing but the tender, evergreen leaves of the sage. Unfortunately, sagebrush is a hated plant in many Western

regions, where cattlemen have used herbicides to clear it from ground they want as pasture, and sage grouse have suffered, especially in the northwest portions of their range.

In the same habitat that supports sage grouse, a patient birder can find horned larks almost everywhere – although spotting these earth-colored birds before they fly can be a challenge. Sage thrashers are another common breeding species from the Canadian border to New Mexico. Smallest of the thrashers, the sage has a relatively short tail and bill, but retains the family's flashy personality; if there are sage thrashers around, rest assured they will make themselves known quickly, zooming across the road and landing, tail cocked, at the top of a sagebrush. Brewer's sparrows and sage sparrows are also both common, with the grayer sage sparrow close to the ground and running.

Birds of the flats

On the wide-open sagebrush flats, birds of prey can be seen for miles. Undoubtedly the most common is the Swainson's hawk, a familiar sight perched on telephone poles and fenceposts, or soaring on pointed wings. A buteo like the red-tailed hawk (which also hunts here), the Swainson's hawk is built on a more delicate plan to hunt smaller prey. At the opposite end of the scale is the powerful ferruginous hawk, a buteo with white underparts and reddish-brown legs that stand out at a distance like a rusty "V." The ferruginous hawk specializes in ground squirrels and big jackrabbits, and is declining over much of its range. Where rocky cliffs and buttes are found, so too will be found the prairie falcon, the common falcon of the West. A bird-eater first and foremost, the prairie falcon takes its prey in midair after a blistering dive, although lizards, ground squirrels and mice will be taken if they are abundant.

ARCTIC TUNDRA

Beyond the boreal forest, where the ground never thaws completely and trees grow in small, besieged pockets if they grow at all, rolls the vast arctic tundra. Covering northern Canada and Alaska, it is wilderness in the truest sense – remote, untouched, and mostly unattainable.

A bird nursery

But the birder who makes the long journey by bush plane finds riches beyond belief, for the tundra is the nursery for many of North America's most beautiful birds. Here, on the lakes and ponds that dot the watery muskeg, live arctic and red-throated loons, gliding serenely beneath the reflections of distant snowcapped mountains. Rarer still is the yellow-billed loon, similar in plumage to the common loon (yet another arctic nester), but with a yellow, rather than black, beak. Much of the birdlife of the arctic is tied to water in one way or another, and for good reason – even though this region receives scant rain, the permafrost that underlies the soil prevents the water from seeping away. The result is an abundance of rivers, streams, lakes, ponds and bogs which are ideal for water birds. Tundra swans and sandhill cranes nest in marshes beside the lakes. White-fronted geese, snow geese, the rare Ross' goose, emperor geese (in Alaska), brant, Canada geese and several species of puddle ducks breed here as well. Four species of eiders – common, king, spectacled and Stellar's – are avidly sought by visiting birders. All are strangely plumaged, but the king perhaps most of all, with a powder-blue head, greenish facial patch and a red bill with a protruding orange frontal shield.

No fewer than 30 species of shorebirds breed in the arctic tundra, although their individual habitat requirements and range restrictions may be more specialized than that. The black turnstone, for

Below This red-throated loon is nesting on the banks of a backwater of the McConnell River in Canada's Northwest Territories. Even while breeding these birds will regularly fly long distances to fish.

Below The spectacular male king eider rises out of the water as part of its courtship display. This arctic duck probably has a mate nearby on a nest in the tundra.

Right Open tundra with dwarf vegetation is the summer habitat of the Lapland longspur. In such bleak surroundings males, like this one, will take advantage of any song post.

example, breeds in salt-grass tundra on the western and southern coasts of Alaska, while the rock sandpiper of the same area chooses its nest sites on grassy tundra on hillsides. Dunlin like soggy tundra near the coast, sanderlings prefer drier upland habitat, and pectoral sandpipers seem content with either.

Breeding in the arctic has some obvious disadvantages – the distance the birds must migrate from their South American wintering grounds, and the limited warm season – but it has some benefits, as well. The 24-hour sunlight of the arctic summer allows an accelerated pace in chick-rearing, and the marshy landscape produces mind-boggling numbers of insects, that make life a torment for visiting humans.

Tundra highlights

Most birders consider the insects a fair trade for all that the arctic offers. Snowy owls, huge and yellow-eyed, watch for lemmings from hillocks; their nests are nothing more than an unlined depression in a small hummock, big enough to hold the eggs. In most years the clutch will only be two or three chicks, but if the lemmings are near

the peak of their population cycle and food is abundant, the owl may lay as many as nine or 10 eggs. Because incubation is started as soon as the first egg is laid, there is a wide disparity in the ages of the siblings. Should food run out, the smallest chicks die first and are often eaten.

Long-tailed jaegers dive-bomb anything that gets too close to their nests, as do pomarine and parasitic jaegers. Where cliffs and river bluffs provide nesting sites, peregrine falcons, rough-legged hawks and gyrfalcons can be found. As justly legendary as the peregrine is, the gyr is awe-inspiring as it towers to the hunt, then spills effortlessly into a dive to capture a willow or rock ptarmigan. In open tundra can be found northern wheatears, arctic warblers and yellow wagtails (in western Alaska), Smith's and Lapland longspurs, common and hoary redpolls and "gray-crowned" rosy finches. Where pockets of forest can be found along river valleys, look for northern shrikes, Bohemian waxwings, orange-crowned and Wilson's warblers, American tree sparrows, fox sparrows and white-winged crossbills. And of course, the possibility of bumping into a bull moose or a grizzly bear only adds to the excitement.

HILLS AND MOUNTAINS: HIGH PEAKS

The golden eagle is the monarch of the continent's rooftop, hunting among the highest peaks of the western mountain chains. Sailing on wings nearly seven feet wide, the eagle nests on remote cliffs, far from humans or the possibility of intrusion.

Found over much of the Northern Hemisphere, in North America the golden eagle is restricted primarily to the West, with a remnant eastern population in the remote mountains of Maine, Labrador and Quebec. A strong bird capable of handling big prey, goldens largely take rabbits, hares and marmots, as well as smaller rodents like ground squirrels, where that prey supply is easiest to catch. Although they have been known to take small foxes and to knock young bighorn sheep off cliffs, their role as predators of domestic sheep has been grossly exaggerated by stockmen.

The golden eagle

Golden eagles defend immense territories, often between 50 and 100 square miles, usually throughout the year instead of just in the breeding season. They apparently mate, if not for life, then at least for many years, reaffirming the pair-bond each spring with thrilling dives and aerial displays. The nest is almost always on a cliff, a large mass of sticks that grows year after year, although some pairs alternate between two or more nests – a strategy that may cut down on parasites and insects, as does the periodic addition of aromatic plants to the nest. Rarely do they lay more than two or three eggs, and as is the case with many raptors, the oldest of the chicks may kill and eat its younger sibling if food runs short. The nestling period is a long one, stretching over 2 1/2 months, for the eaglets need time to develop their muscles and coordination. Even after they leave the nest, they remain dependent on their parents for many more weeks as they laboriously learn to hunt.

Rabbits and hares rely on speed to escape an attacking eagle, but the hoary marmots that live in the high country cannot run that well. Built like

Right This fully mature female golden eagle is seen at its huge nest located on a sheer cliff protected from wind and rain.

Above These ravens display the large head and bill, together with the wedge-shaped tail, that distinguishes them from other black crows. Their aerial mastery is ideally suited to a life among the highest of mountains.

Right The rosy finch is a bird of the high western mountain tops. This black form, marked by a pale gray nape, is found in the central basin of the Rocky Mountains.

the groundhogs to which they are related, hoary marmots may weigh up to 20 pounds, and live in colonies among the talus slopes and rock slides above the timber line. When they feed they stay close to the entrances of their burrows, one eye always cocked skyward for danger. Should an eagle appear, sharp alarm whistles echo over the rocks, and the marmots scramble for cover.

The birdlife of the peaks is sparse. Ravens live here, nesting on cliffs and feeding on carrion and small animals; in many ways, these corvids function ecologically as raptors, even though they are not related to hawks or eagles. At the edge of the snowfields, where only a scattering of scraggly grass and brush can survive, are flocks of rosy finches. There are four main color phases, once thought to be separate species. The "gray-crowned" phase is the most widespread, occurring from Alaska down through California at high altitudes; the "black" is restricted to the Rockies and adjoining ranges; the "Hepburn's" rosy finch is in coastal ranges from the Aleutians to the Pacific Northwest, while the "brown-capped" is found only in the central Rockies. Whatever the color phase, the rosy finch is able to eke out a living in the least hospitable of circumstances, nesting in rock crevices or cliff walls where eggs are partially shielded from the biting wind. In winter, large flocks roost in caves and buildings.

ALPINE TUNDRA

Mountains make their own climate, and their very size creates bands of different living conditions on their slopes which change with the increasing altitude. In Wyoming and Montana, for example, the foothills are predominantly sagebrush and grasslands, up to an elevation of about 7,500 feet. For the next 1,000 feet, to a cooler elevation of roughly 8,500 feet, are forests of Douglas fir and aspens, which are succeeded by lodgepole pine, then subalpine fir and Engelmann spruce. The highest of the mountain life zones, above the timberline, is the alpine tundra. In terms of growing season, temperature, and plant life, the tundra here is very similar to that of the arctic.

The camouflaged ptarmigan

Like the high peaks, the birdlife here is not rich, either in diversity of species or numbers of individuals. Golden eagles regularly hunt here for marmots and ground squirrels, retreating to cliff faces to nest. Ravens are also conspicuous, as are rosy finches. Common but harder to see are white-tailed ptarmigan. The most southerly of the three ptarmigans, the white-tailed is found in alpine tundra from Alaska and British Columbia down to New Mexico, although birders must look closely to see them, for the ptarmigan has raised camouflage to its highest degree. In winter, both sexes are pure white, betrayed against the snow only by their black beak and eye, and by a thin comb of red flesh above the eye. But as the spring approaches, in late May and June, the ptarmigan molt into their summer plumage. Even as the land becomes splotched with bare ground and melting snow, the ptarmigan takes on a piebald appearance, with white wings and tail feathers. Against the stony ground, the match is perfect, and a human may walk within feet of a motionless bird without detecting its presence.

Unlike many lowland grouse, the white-tailed ptarmigan has no elaborate courtship dance; the male calls and struts slowly to win his mate's affections. The nest is a shallow depression in the tundra, lined with grass and leaves and holding up

Below Smallest of North American birds, the calliope hummingbird is a summer visitor to high alpine areas, as well as canyons and mountain streams.

to 7 or 8 eggs, which the female incubates alone. To avoid detection, she will often swallow any white feathers that she molts, although in some nests feathers are woven into the lining. The chicks are precocial, leaving the nest a short time after hatching and able to forage for insects and tender buds themselves. The female watches them closely, however, and will put on a convincing distraction display should a predator (or birder) come too close.

Alpine songbirds

There are a few songbirds, beside rosy finches, in the alpine region. The water pipit, more familiar as a wintering bird on beaches and in agricultural areas, nests in both alpine and arctic tundra; in the mountains its sky-song, in which the male circles high and then drops slowly while singing, is a common sound of spring. The white-crowned sparrow, abundant in many western habitats, is also common in alpine regions. The calliope hummingbird, the smallest bird in North America, is more common in forest meadows on the mountain slopes, but will range above the timber line when the many species of alpine flowers are blooming.

Above High mountains in high latitudes are the haunt of the white-tailed ptarmigan seen here in summer plumage. The mottled grays merge well with lichen-covered rocky screes and plateaus.

Right In winter the white-tailed ptarmigan adopts an all-white plumage as camouflage among the snow covered mountain slopes that it inhabits. The male has a bold red comb.

ALPINE FORESTS

Above the sagebrush and below the alpine tundra are the forest life zones mentioned earlier. These are coniferous forests, but the species of trees, as well as the birds that live among them, are so vastly different from the boreal forest of Canada and New England, or the pinewoods of the South, that they bear separate discussion.

Although the trees grow in zones segregated by altitude, the birds, being mobile, are not quite so closely tied to a particular band. Compared to the alpine environment, the pickings for the birder are much more rewarding, although the dense growth can make seeing the birds difficult. Some, like the goshawk, are by nature retiring, and it is only by sheer luck that one's path crosses this spectral raptor's. Sharp-shinned hawks are more common and more frequently seen, especially in the morning, when these accipiters usually enjoy soaring above the trees with their flap-flap-flap-

Left Clark's nutcracker is often common in the conifer forests of the Rockies, where it lives in noisy flocks. In summer it frequently visits camp and picnic sites for leftover food.

Above The mountain chickadee is a common inhabitant of high conifer forests throughout the Rocky Mountain system. The thin white eyebrow is a useful field mark.

glide flight pattern. On the ground, look for ruffed grouse (often the silver color phase, rather than the reddish-brown found in the East) and the beautiful blue grouse, which replaces the similar spruce grouse in most western mountains; the "Franklin's" spruce grouse, a distinctive subspecies, is found in the northern Rockies and Cascades, however.

Stellar's jays, like blue jays with black heads, flock in noisy aggregation, especially around campsites where food may be stolen from an unwatched picnic table. Even bolder is the gray jay, which will swoop down and land on one's hand for a meal; such behavior should not be encouraged, however, because human food is rarely nutritious for wild animals. Camp cooks can be forgiven for feeling put-upon should yet another mooching corvid arrive – the Clark's nutcracker, a crestless jay with a gray body and flashy black-and-white wings.

The great gray owl

If there is an ultimate bird in the mountain forests, one for which all visiting birders hope, it is surely the great gray owl – "the gray ghost" to those who have sought for it and failed. In length it is by far the biggest of North America's owls, although the bulk is mostly feather and it is out-weighed by the much stronger great horned and snowy owls. Huge facial disks, marked with concentric circles, frame two yellow eyes; the rest of the bird is a cryptic jumble of gray, black and white streaks. For all its size, the great gray is a fairly weak hunter, rarely taking prey much larger than mice and voles, often captured at twilight when a lucky observer can thrill to the owl's silent flight. Great grays are uncommon residents of the northern Rockies and Cascades, becoming somewhat more common the further north into their range one goes. But they are never obvious, and the sight of one sitting quietly in a fir tree is the birding event of a lifetime.

Smaller birds

Visitors will find an interesting variety of smaller birds in the alpine forest. Mountain chickadees, red-breasted nuthatches, hermit thrushes, Townsend's solitaires, western tanagers and Cassin's finches are common among the conifers themselves, while in the scattered thickets of aspen and other deciduous trees, watch for orange-crowned, MacGillivray's and Wilson's warblers, Lincoln's sparrows, Swainson's thrushes (especially near water), warbling vireos, mountain bluebirds and white-crowned sparrows. In wildflower-studded meadows, calliope, broad-tailed and rufous hummingbirds stake out feeding territories, vigorously chasing away other hummers and even nectar-feeding insects like butterflies. Olive-sided flycatchers sing their clear *Hic! Three beers* song from conifers, a habitat they share with Hammond's and western flycatchers, two of the confusing Empidonax group.

Below The male blue grouse makes much use of its inflatable, bare neck sacs in display. The tail is fanned, the head combs raised and the bird frequently flutters into the air to impress potential partners.

FROM BIRDER TO ORNITHOLOGIST: ETHOLOGY

Above According to James Fisher an increase in fish waste and decrease in persecution may be behind the establishment of new fulmar colonies along many coasts.

Right The song sparrow was the subject of an in-depth study by Mrs. M.M. Nice that set the style for all the behavioral monographs that followed.

At first we are content to identify the birds we see, to watch and enjoy them and become familiar with their everyday lives. Gradually we seek to know more. We dip into books, subscribe to magazines and journals, attend lectures, join societies and clubs - but most of all we begin to watch birds in a different way. Our watching becomes purposeful. Of course, such watching needs a background in ornithology to understand what things to look for, but an intensive course of reading together with regular field trips in the company of other birdwatchers soon gets most of us going on research of some sort.

"Research" is a formidable word. It smacks of sterilized laboratories and has a musty, old book smell to it. In many areas of knowledge, such as physics and chemistry, research is the preserve of white-coated Ph.D.'s backed by the resources of universities or international companies. With birds it is different. If bird watchers want to study some aspect of the lives of birds, they can go ahead and do so. Many of the most important elements of birds' lives have been described by amateurs working on their own. A teacher called David Lack could not stop looking out of his classroom window to watch the color-ringed European robins that haunted the school grounds. The result was the classic *The Life of the Robin*, and David Lack became director of a famous ornithological institution, the Edward Grey Institute in Oxford. While there may now be more professional ornithologists than in Lack's day, there are still more birds than there are professionals to cover them. Thus research is full of possibilities for amateurs to make meaningful contributions.

Research may be individual or organized, and may consist of field work or museum and library study. Some types of research lend themselves to field work – the study of breeding behavior, for example – whereas others concerned, let us say

with the distribution of a species, are essentially researches through published material. It is just not possible to learn where a bird is found by searching for it: one must rely to a large extent on the efforts of others. But this does not make such research unproductive. There are untold amounts of information buried in miscellaneous journals which could be collated and published.

Bird Psychology

Ethology, the study of animal behavior, is often thought of as the preserve of the psychologist. It is not. Bird behavior can be watched and noted by anyone with eyes to see. Edmund Selous produced some of the most important bird books published this century, summarizing his observations of

Above German ethologist Konrad Lorenz imprinted greylag goslings, as their surrogate parent, thus gaining insight into a hitherto unknown area of animal behavior.

Below The European robin proved the ideal subject for a youthful David Lack to study in depth and produce a classic bird book. Work on common birds produces faster results than studies of rare or elusive species.

many common European species. Anyone who would follow in his steps should read his books. Indeed, to study bird behavior it is necessary to read what others have achieved before you if you want to do anything of any significance. But it most definitely is not neccessary to have reached degree standard.

If bird behavior interests you, study it. Watch carefully and record exactly what you see. Omit nothing, for the very detail that you require may only be apparent in retrospect. Try to record as much as possible, for too much data is certainly better than too little.

If your ideas of research seem to lead towards the romantic desert island approach, forget it. Ethology benefits most from work on birds that are common and easy to watch near the observer's home. Indeed, seabirds are the best subjects for case studies simply because large numbers are concentrated in small, easy-to-work locations, and because they are generally easy to see and comparatively tame. The observer can watch a great many individuals in a very short space of time. So never try to study the behavior of a rare or elusive species when a common and tame one would do as well. There are still plenty of subjects only a short walk away.

CENSUSES

If the study of bird behavior is essentially an individual activity, inquiries into distribution are clearly cooperative affairs. In Britain the British Trust for Ornithology (BTO) organized the field work and production of an *Atlas of Breeding Birds*, an idea that has been taken up in many other parts of Europe and a number of U.S. states. The success and popularity of this venture prompted the BTO to follow with a winter atlas project and it presently plans a repeat of the breeding atlas.

Based on the National Grid 10km squares, into which Britain is divided, bird-watchers volunteered to search systematically for breeding birds in the square nearest their home over a period of several years. The result has been an incredible increase in our knowledge of birds in Britain. It was the element of competition, with one's own record as much as with others, perhaps, that made for such success. There was always the chance of another bird, new for the square, being found in a seldom-visited part of the allotted zone.

As a consequence, habitats were searched that had not been visited by ornithologists before. Spotted crakes, formerly considered rare and irregular breeders, but ones that are so easy to overlook, were found here and there throughout southern England. Other easy-to-miss species were also found to be much more widespread than had been appreciated before.

The great bird count
The annual Audubon Society Christmas Bird Count, started 90 years ago, now encompasses all U.S. states and Canadian provinces, as well as Guam, Central America and the Caribbean basin. Each year, more than 40,000 birders fan out to comb 15-mile-wide circles, recording every species

Below By making six visits to an area and plotting the position of each singing robin on each occasion, an ornithologist is able to build up a picture of the total number of singing males and, therefore, of the number of breeding pairs.

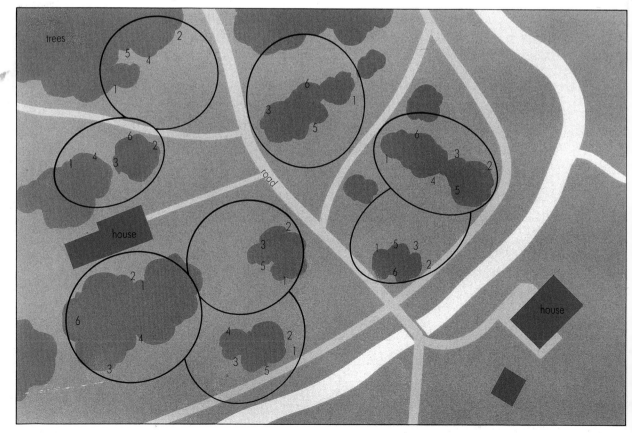

Left Counts of seabirds, such as the gannets on Scotland's Bass Rock, enable ornithologists to monitor the populations of these highly gregarious birds.

Below The distribution of gannets, evenly spaced across the rock, indicates that each bird is on its nest. Non-nesting birds are spaced irregularly.

of bird and every individual they find in a single, 24-hour period around Christmas. Up to 120 million birds are counted each year, and the results (800 or 900 pages of them) are published annually in *American Birds*, the Society's professional journal. While the count itself is an excuse for birders to indulge their listing instincts and the spirit of competition, the results have been invaluable in tracking long-term population trends in many different species.

While it is possible (albeit difficult and time consuming) to count all the colonial seabirds, it is clearly beyond the scope of anyone to count many land-based birds. Clearly there will be more at the end of the breeding season than at the beginning. If it has been a good breeding season, there will be more birds than if the weather has been foul. Thus, the only significant number from a population angle is the number of pairs that commence breeding.

There is no need to count all the birds but just to keep an eye on populations. Awareness of falling populations really started back in the 1950s when many watchers noted declines in the numbers of what had been, until then, quite common birds. Gradually the decline began to be linked with the use of agricultural chemicals used as insecticides and seed dressings (the chlorinated hydrocarbons). These persistent chemicals eventually found their way into the body tissue of birds of prey: there was a catastrophic decline in the number of peregrines. As a result effective preventive measures were taken just in time.

Frightened that they had not known about the effects of change on bird populations, and anxious that a similar disaster should be recognized earlier, the BTO introduced the Common Bird Census to monitor changes in the population of commonplace birds.

POPULATIONS

Above These bald eagles in Glacier National Park have gathered to feed on the fall run of salmon.

Right Peregrines were one of the primary victims of pesticide poisoning during the 1960s. They are still absent from huge areas of their former range.

In the Common Bird Census system, an area of 40 hectares or so of farm or woodland is selected and visited by an observer five or more times during the peak of spring activity. On each visit the exact position of each singing male bird is plotted on a map. At the end of the season, when all the summer migrants, and the resident birds, have settled down to breed, the maps for each visit are laid over one another and a single map drawn of the completed season's work. From this it is possible to see that the various individual birds have maintained their territories, for records of singing males in different months form clusters in the final map. These clusters are counted and the number of singing males of each species totaled for the year.

While the Common Bird Census does not count every bird, it does give an accurate idea of whether a particular species is doing well or not. CBC workers are scattered throughout the country covering every type of habitat. Some have been covering the same area for 10 or more years and have enjoyed every moment of their constructive field work as well as the fascination of compiling maps and drawing together the final results.

Building up a picture

When a picture for the whole country emerges, we can see the remarkable recovery that birds can make after an exceptionally hard winter has decimated their numbers. When observatories reported fewer whitethroats in spring, CBC workers were asked to forward their results for the species as a matter of urgency. Within a few weeks it was apparent that there had been a disaster - the numbers of breeding whitethroats had suffered a catastrophic decline. Looking for a cause revealed that the wintering grounds of this attractive little warbler had suffered a major drought. Soon newspapers and other media were full of stories of human famine in the same area, the Sahel region south of the Sahara Desert. Pictures and film of starving people, and of children in particular, caught the conscience of the western world and led to massive fund raising events such as "Band Aid," "Live Aid," and so on. Yet the work of amateur bird-watchers could have been used to draw attention to the drought months before it actually hit the headlines and much suffering could have been avoided.

The populations of birds can, then, be

monitored in various different ways. Widespread and common birds can be sampled. Birds that are concentrated at a few major sites to breed can be counted, but so too can birds that, while they may breed scattered over huge areas, concentrate into well-defined areas at other times of the year. Wildfowl are particularly concentrated during the winter and with several species of geese it is possible to count the complete world total. There are, for example, some 75,000 barnacle geese in the world with three quite distinct populations wintering in northwestern Europe. Fluctuations in numbers, as well as the percentage of young birds in the flocks, can tell us which population has enjoyed a good breeding season and which a poor one. We can counter the arguments of sportsmen and farmers who wish to hunt the geese and we are always alert to disaster.

Censusing birds provides one of the most valuable tools a conservationist can have; it is primary research, often by amateurs, that provides the basis.

Right Barnacle geese, once in serious decline due to overshooting, have responded to protection and increased dramatically during the past 30 years. Their concentration at a few major wintering sites makes them particularly prone to disturbance.

BANDING

The study of birds by individually marking them with bands is comparatively recent. Yet in the space of 70 years it has led to a complete revolution in our understanding of their lives. Migrant birds have been proved to fly vast distances to and from regular summer and winter homes and even to stop off at the same places along the way. We now know exactly how long an individual wild bird lives. We can estimate the population of a species by comparing the rate at which banded birds are retrapped in a particular area with the overall chances of retrapping. If, as often happens, the same bird is retrapped in the same area, then the same measurements can tell us whether or not the bird is gaining weight and how far it has progressed with its molt. We can find out how old a bird is when it first starts to breed and much more besides.

Taking wild birds for banding requires a license from the government, and to obtain a permit is a long and painstaking job. Would-be banders must be trained by existing banders and serve a sort of apprenticeship. They must learn how to catch birds safely, how to handle them, how to put bands on and keep accurate records. They must learn the rules and regulations concerning what birds they may band. So the starting point is to contact a local bander or visit a bird observatory.

Mist nets

Today most birds are caught in mist nets - fine black terylene nets that disappear when suspended against a dark background. These nets must be carefully sited (they can catch cows and people too) and visited every 15 minutes or so.

Above Heligoland traps, like this one on the island of Lundy, England, are simply giant wire-netting funnels placed over suitable cover to trap migrant birds for banding.

Left Mallard, caught by cannon-netting are transferred to holding pens prior to being banded in South Dakota.

Birds fly into a mist net and drop into a pocket of netting below a shelf-string. There the majority lie still until they are extracted by the bander - not a job for the clumsy or quick-tempered. Some species fight furiously to escape, grabbing footfuls of netting from all directions. The longer they are left, the longer they take to extract. Sometimes a chickadee will tie itself in a virtual ball of netting and pose a considerable problem for the bander. In extreme cases, the bander must be prepared to cut

Below Banding Manx shearwaters has proved one of the most productive forms of migration study. Individuals transferred from Britain to the U.S. have returned to their nests in a matter of days.

Right This trumpeter swan is being returned to the water at Red Rock Wildlife Refuge, Montana, with a colorful neck band boldly coded so that it can be read without recatching the bird.

the net to extricate a bird safely.

The trapped bird is marked with a light aluminum band crimped loosely on the leg. The band is individually numbered, and carries an abbreviated address for the U.S. Fish and Wildlife Service in Washington D.C., which serves as clearinghouse for North American band recoveries. Should you find a banded bird, send the band number, along with where, when and how the bird was found, to the address on the band. Later, you'll receive a certificate detailing where the bird was banded, when and by whom.

Despite expressed fears, banding, when carried out by an expert, does no harm to the bird. If it did, the whole point of it would be destroyed, for the aim of banding is to study the normal behavior of normal birds. Accidents do happen, but they are very few indeed, and the gains in knowledge far

outweigh the tiny element of risk. In fact, banding is now regarded as a primary element in conservation work and many bird organizations actively pursue programs of banding in their efforts to monitor populations and protect birds. There is clearly little point in spending scarce resources on protecting a bird on its breeding grounds if the greatest danger lies thousands of miles away in its winter quarters.

The value of banding

At one time banders simply banded; but now they take full advantage of their opportunities to extract the maximum amount of data from the living bird in the hand. Weighing is particularly important. During the day a bird's weight varies considerably, but prior to migrating birds put on fuel in the form of body fat; some species may double their weight.

A great deal of what we know about the lives of birds is a direct result of banding. By marking an individual bird in this way we can, via a recovery or by others retrapping the bird, learn where it breeds, where it winters, and about the route taken between the two. We may learn something of the speed with which it makes its journeys and of places where it stops-over to feed and recuperate. We have learned, for example, that, although barn swallows winter throughout Africa south of the Sahara, British swallows spend their winter in South Africa, while German swallows winter in West Africa. We have discovered that, while birds migrate on broad fronts, crossing seas and deserts along the way, there is still nevertheless a "migrational divide" in Europe. Some birds move southwest, while others move southeast. We explain this in terms of the direction of colonization following the last ice age.

Building up a map

Most banding stations and banding groups build up sufficient recoveries of birds to justify creating a map for particular species. Pinned to the wall of the laboratory or banding hut, these show not only the destinations of birds, but also their origins and the route between the two. The organizers of national and international banding projects produce even more records and their maps show a virtually definitive picture of the movements of individual species.

A map of recoveries of British-banded white-fronted geese, for example, tells us that the birds that winter in England come from the population that breeds along the coast of northern Russia and the islands of Novaya Zemlya. But it also shows that these birds do not follow a direct route, but make a loop migration southward to central Russia on their way westwards to northern Germany, Holland and the Channel coast of France.

A similar map for knot shows that birds banded in Britain have actually made a huge transatlantic flight from western Greenland and the Canadian Arctic and that many of these birds continue along

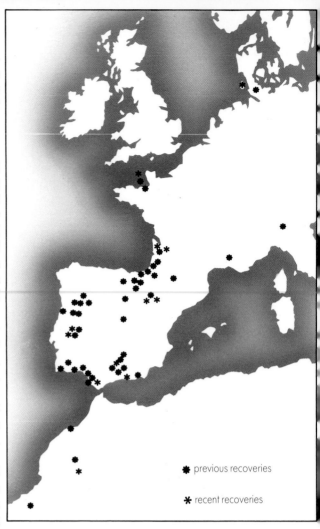

✳ previous recoveries

✱ recent recoveries

Above Recoveries of British banded robins show a general movement to the southwest, with most recoveries in France, Portugal, and Spain.

Above right Banding ferruginous hawks in North Dakota produced recoveries that show a general southward movement, though some wander a little to the north.

Left An ornithologist in a field laboratory, or banding hut, is banding a sanderling, one of the long distance migrants that banders favor.

to-west across the path of birds flying between Europe and Africa. In the Americas the physical barriers run largely north to south, roughly parallel to the routes taken by birds. This has had the effect of concentrating birds into particular "flyways" in North America, whereas European birds have no advantage in concentrating in this way. Such an overview would be impossible without the backing evidence of banding.

• recovered during first year of life

• recovered during subsequent years

Below Banding some species, such as this Cassin's auklet in Alaska, may be unrewarding from a recovery standpoint, but it does offer ornithologists the opportunity to handle live birds and take weights and measurements.

Bottom Banding birds requires skill, training, and special equipment like these pliers being used to band a migrant European blackcap. ornithologists the opportunity to handle live birds and take weights and measurements.

the coasts of the southern North Sea and the Atlantic as far as Spain and Morocco. More typical is the map of British banded European robins, which shows that robins banded in Britain move Southwestwards to winter in western France and south and west Iberia.

In North America banding has had similar dramatic results. Who would have guessed at the number of birds that regularly migrate across the Caribbean rather than follow the land route through Mexico and Panama? A map of the banding results for the bobolink not only shows this route, but it demonstrates that bobolinks from the western U.S. head eastward to join it. In fact this bird has spread westward across the U.S. during the present century and western colonizers are following the colonizing route for the first part of their migration.

In the Euro-African migration system, most of the significant barriers to bird movement run east-

MIGRATION STUDIES

It will by now be apparent that I have a particular penchant for the migrations and movements of birds and a fascination with identification. Not surprisingly these two aspects of birding go hand in hand, and I am far from being alone in such enthusiasms. Watching migration is, in some instances, as dramatic a spectacle as the wonderful concentrations of birds of prey at Hawk Mountain in Pennsylvania. In Europe there are similar spectacles at the Straits of Gibraltar, the Bosphorus and at Falsterbö in southern Sweden. At Gibraltar and the Bosphorus there are also great spiraling flocks of storks, although at a different season. These are spectacular migrations that anyone can go and watch - bird movements at their most obvious and dramatic.

Most other birds are less concentrated and their migrations are thus more difficult to detect. We can see birds appear in spring in our yards and know that they have just arrived. We can watch a flock of birds at a favored site and know they are migrants because they were not there yesterday. But real dramas are hard to come by and usually require a visit to a special migration watchpoint. In North America there are favored points along the Gulf Coast and on the northern shores of the Great Lakes such as Point Pelee and Long Point. In Europe most migration work has been done in Britain and along the southern shore of the Baltic.

Migration watch points

Britain boasts a fine network of bird observatories on islands and headlands along its extensive coastline. These are all excellent places to watch

Left This migrant house martin has become lost after dark and has taken refuge near a lighthouse, doubtless attracted by the beams of light.

Above White storks migrate in large flocks that concentrate into huge, wheeling masses at the narrow sea crossings of Gibraltar and the Bosphorus.

migration, to seek out migrants and note the daily comings and goings of a large range of species. Many of these watch points have nearby, or adjacent, lighthouses and it was the birds attracted to the lights that first drew the attention of observatory founders to their potential. Many birds died as a result of flying into the lights, and picking up and identifying the corpses of nocturnal migrants was, at one time, an integral part of the work of observatory enthusiasts. Today, these dangerous structures are fully illuminated so that even dazzled birds can see them and take avoiding action. In America, birds frequently collide with radio and television towers and several workers pick up and identify the bodies on a daily basis. While most bird observatories are banding stations, they also attempt to note every bird seen

Left Migrants often get lost when bad weather prevents accurate navigation. This wheatear has taken refuge on a ship 200 miles off the French coast, where it will await better weather before reorientating and continuing its journey.

within the observatory area each day. Particular attention is paid to patches of scrub where migrating birds seek shelter. Mostly this is a matter of "doing the rounds" early every morning when birds are most active. Waiting patiently on the sunny and sheltered side of cover is far more effective than thrashing around in the hope of disturbing birds; and it is often surprising what the more patient birder can find. The number of birds will vary day by day and the casual visitor will gain little impression of migration from a single visit. It is regular daily watching which brings to light any changes and which indicates what birds are passing through.

Just occasionally, particular weather conditions will produce a fall-out of migrants that it is impossible to ignore. Where there were a few, there can quite suddenly be hundreds or even thousands of birds. Such falls are a dramatic sign of migration, an event which usually passes largely unseen, and it is every birder's ambition to experience such a fall. Inevitably it is usually the resident wardens at observatories that experience

such dramas, together with the few birders who, by sheer good fortune, happen to have chosen that particular day for an outing.

In my early birding days I remember vividly poring over the daily weather forecasts watching for what I thought would be perfect "fall" conditions. What I was seeking was high pressure over Scandinavia which would provide perfect migration weather for nocturnal migrants to set off over the North *Sea en masse* towards Britain. I then needed a deep depression to arrive in southeastern England with rain and cloud to force the birds down. Suddenly the map predicted everything I wanted. The problem was where to go? The front associated with the depression was predicted to pass through Kent at dawn. I arrived early in pouring rain and had a miserable birdless day. I returned home to be informed that the Suffolk coast had enjoyed the most dramatic fall-out in history. I'd got the day right, but the front had moved through too fast and the birds had been disorientated over the southern North Sea instead of over Kent.

CONSERVATION

Being a birder does not inevitably mean that one is a conservationist, in the same way as being a conservationist does not turn one into a birder. Nevertheless, it is difficult to watch birds without being interested in their future and all birders, bird-watchers, and listers should, at least, be members of their national and/or local bird conservation organization.

National organizations

Just as birding is becoming progressively more international, so too is bird conservation. Local conservation societies have local aims, national ones national aims, but birds are the most mobile animals on Earth and their conservation has to be approached on an international scale. By and large international organizations are low on both funds and muscle, and many of the most important measures are originated by go-ahead "national" organizations. Prime examples are the work of the National Audubon Society and the British Royal Society for the Protection of Birds. Both organizations are primarily concerned with the birds of their own country, but both recognize the international implications of bird conservation. There are, of course, fully international organizations such as the World Wide Fund for Nature (WWF), the International Council for Bird Preservation (ICBP), and the International Union for the Conservation of Nature (IUCN); but all have great demands on their limited funds and can barely scratch the surface of bird conservation.

Specialist organizations

There are several more specialized organizations that attempt to deal with specific areas of bird conservation on a world-wide scale. Primary among these is the Wildfowl Trust, the brain-child of Peter Scott, and a model that has shown the way that other organizations could go. The Trust is actively involved in wildfowl conservation throughout the world and its Hawaiian Goose project has been successful in saving, studying and releasing this endangered species back to its native land.

Surprisingly few other specialized organizations have followed the lead shown by the Wildfowl Trust, although it has spawned many imitators

Above The Californian condor is all but extinct. In a last desperate bid to save this bird, all the remaining wild birds were caught to become part of a captive breeding program.

among duck and goose enthusiasts. There are, however, organizations that specialize in the conservation of pheasants, birds of prey, owls and cranes. The latter is something of a model of its type and owes its existence to the enthusiasm of two young Americans - George Archibald and the late Ron Sauey. Based in Wisconsin, the International Crane Foundation has a collection that includes every species of this spectacular group of birds. It is active in the study of cranes and has bridged political boundaries in its conservation work. If anyone has the odd few million (pounds, dollars, deutschmarks or yen) looking for a home I would recommend it be sent instantly to George.

Left The peregrine captive breeding program at Jackson Hole, Wyoming, is just one such attempt to restore the falcon to its previous numbers after the massive pesticide kill of the 1960s.

Above Conservation work takes a variety of forms, but ornithologists are progressively becoming land management, rather than species protection, oriented. Over-grazing, seen on one side of this fence, can ruin perfectly good grassland.

There are other bird groups that have attracted the attention of small bands of enthusiasts for a variety of different reasons. Peregrines have been bred to replace the wild populations wiped out in a number of areas by the chemical disasters of the 1960s. In many cases this is, however, no more than enlightened self-interest by keen, perhaps obsessive, falconers. Similarly, there are captive breeding projects devoted to bustards and backed by enormous funds in the Middle East. Having exterminated their own bustards as prey for their falcons, the sheiks are having to look elsewhere, or breed their own. It would be good to know that the resources available were to be used for the international conservation of this generally neglected group of birds . . . but I doubt it will.

CONSERVATION PROJECTS

Conservation projects come in all shapes and sizes. Erecting a garden nest box is as much a bird conservation project as creating a 500 acre reserve. Yet what an individual can do must inevitably be limited when compared to a large and influential organization. Governments, local and state authorities, multinationals and giant corporations, rich and powerful philanthropists can all contribute by using their resources to establish reserves and bird-friendly policies - but what can be done, can also be undone. Land ownership by conservation bodies is the key.

A hundred years ago a nature reserve was an area with a fence around it and notices saying "Keep Out". Today's reserves are actively managed to produce the right conditions for the species most in need of help. Land can be converted to marshes, to swamps, to forests and even be farmed specifically to provide food for wild birds, but it will also be used to educate and enthuse visitors. Of course if you can begin with an area that is more or less what you want, you start with immense advantages. But even if the newly-acquired reserve is poor, as long as the potentiality

is there it can be transformed.

One of the best examples of this dynamic approach to bird habitats is to be found at the RSPB's Minsmere Reserve in Suffolk, England. Here, with the help of the Army, a series of shallow coastal lagoons have been extended to create "The Scrape", an artificial bird breeding and feeding site of immense importance. The lagoons existed before the Army moved in, and sheltered a few passage waders: the rest is artificial. Islands covered with polythene (to prevent weeds choking them) have been created and the sheeting covered with pebbles. Here, avocets breed alongside common and sandwich terns.

These habitat-creating activities are dramatic, but control of habitat - clearing thickets and overgrown pools, controlling water levels to preserve unique reedbeds, constructing artificial

Below left Active conservation often involves strenuous maintenance. These volunteers are busily planting trees in an area that has been previously felled, in an effort to attract different species of birds.

Left Whooping cranes have been the subject of one of the most intensive conservation projects ever. After years of slow decline, a captive breeding and egg translocation program has established a secondary population in the U.S. These birds are at their traditional wintering zone in Aransas.

breeding sites, felling, blowing up as well as planting trees - are part of what nature reserve maintenance is all about.

Species protection

Conservation does, however, also concern itself with species protection, especially with migrant birds that may need a chain of sanctuaries where they can find refuge. The whooping crane of North America seemed doomed to extinction despite all the efforts made to protect the last two or three dozen birds. Nesting among the wild marshes of Canada's Wood Buffalo National Park, where protection activities would probably do more harm than good, these splendid birds then made a length-of-the-continent migration to winter along the Gulf Coast at the Aransas Sanctuary in Texas. Over the years the migrating birds were accompanied by watchful aircraft to ensure that they were not shot up by misguided sportsmen. Yet, year after year, numbers see-sawed around in a steadily declining population. In desperation, a captive breeding flock was established and a new population created. At the time of writing the gamble would appear to have paid off and the number of cranes seems to be increasing.

In Britain, the red kite was once down to a handful of birds based in the remote valleys of central Wales. Slowly and painfully the number has increased by dint of the devoted work in nest guarding of a small band of enthusiasts. Today the kite would seem secure, but still every year nests are robbed of their eggs.

These are just two examples of what has been achieved in two different continents. There is no end to what needs to be done.

GLOSSARY

Adaption: the way in which a bird has changed to meet the circumstances in which it finds itself.

Adult: a bird that has acquired its full plumage. To be compared with juvenile or immature plumages.

Aquatic: a life-style associated with water.

Avian: relating to birds.

Axilliaries: a group of feathers located where the underwing meets the body - the "armpits".

Bar: a mark across a feather or group of feathers (see Stripe).

Belly: underparts of a bird between breast and undertail coverts.

Bend of the wing: the point where the wing changes from extending forward to backward.

Cere: a bare patch at the base of the bill in such groups as birds of prey and parrots.

Clutch: the eggs of a bird in its nest.

Coniferous: cone-bearing trees, usually evergreen.

Conspecific: where two (or more) differing birds belong to one species.

Coverts: groups of small feathers that cover the base of the major flight feathers, e.g. wing coverts, tail coverts.

Crepuscular: active at dawn and dusk.

Crest: extended feathers on crown.

Deciduous: broad-leafed trees that shed their leaves.

Decurved: down-curved.

Emarginated: wing feather that narrows on the outer edge.

Ethology: study of behavior in the natural environment.

Extinct: no longer in existence.

Feral: domestic forms that have gone wild.

Field Guide: a book of bird identification for a specific area.

Flanks: sides of the body below the folded wings.

Frontal: on the forehead.

Gamebirds: birds hunted for sport or food.

Gregarious: birds that form flocks.

Gular: throat.

Insectivorous: birds that eat insects.

Juvenile: the plumage in which young birds leave the nest.

Lores: area between the bill and the eye.

Mandibles: the two parts of a bird's bill.

Molt: the process of shedding feathers and replacing them.

Nape: hind neck.

Nocturnal: active at night.

Omnivorous: birds that eat a wide range of foods.

Oologist: egg collector.

Pelagic: birds that live in the open ocean.

Phase: plumage differences within a single species.

Preen: a bird's method of feather care and maintenance with its bill.

Primaries: the outer flight feathers that act as the means of propulsion.

Race: sub-species.

Raptor: bird of prey.

Rump: area of body above the tail.

Scapulars: feathers that cover the area where the upperwing joins the body.

Secondaries: group of inner flight feathers responsible for lift.

Shorebird: wader; term usually used in North America.

Species: a group of birds that are capable of interbreeding to produce fertile young and which do not usually breed with others.

Speculum: patch on upper wing of ducks.

Stripe: lengthwise mark on feather or group of feathers (see Bar).

Subspecies: a group of birds within a species that can be distinguished from other groups.

Tarsus: leg bone immediately above the foot.

Tertials: wing feathers between secondaries and body.

Thermal: rising column of warm air; used by soaring birds.

Vernacular: English name as opposed to scientific name.

Wader: term used in Britain for shorebird; in North America, refers to herons, egrets and ibis.

Wildfowl: ducks, geese and swans.

INDEX

page numbers in **bold** refer to illustrations

A

accidentals 72, 75
albatrosses 20, 23, 51, 55, **55**
  black-browed 110
  Laysan **21**
  yellow-nosed 110
alpine forests and tundra 136-9
anatomy **18**, **19**, 24-7
anis
  groove-billed 125
  smooth-billed 125
arctic tundra 132-3
Audubon Society 80, 142-3, 152
auklets 100
  Cassin's 149
  least 100
auks 19, 86
  great 19, 100
  northern 19
avocets **26**, 154

B

banding 146-9
beaches and dunes 102-3
bee-eaters 21
bee hummers, Cuban 48
binoculars 56-7, **56**
bird behavior 40-3, **40-3**, 141
bird names 38-9
bird song and calls 34-7, **34-7**
bird tables **12**
bird watching see birding
birding 7-17, 60-1
  at dumps 80-1
  at sewage ponds 82-3
  bird calls 36-7
  bird gardening 16-17
  description of birds 66-7 ,
  equipment 60-1, **60-1**
  feeding birds 12-15, **13-15**
  identification of birds 38-9,
    62-3, 64-5
  in cities 76-7
  in parks 84-5
  in suburbs 78-9
  photography 68-9
  provision of nest boxes 11
  recording **37**, 70-1
birding-for-charity 72-3
birds
  accidentals 72, 75

anatomy **18**, **19**, 24-7
arrivals 74
bills 24-7, **24-6**, 63
breeding 28
  eggs 48-9, **48-9**, 50-3
  feathers 22-3, **22-3**
  feeding for migration 29
  feeding the young 51, 52
  feet and legs 24-5, **27**
  flight 18-23, **18-23**
  migration 30-3, 74-5, 102-3, 146-51
  names 38-9
  plumage 20-1, 22-3, **22-3**, 42-3,
    42-3
  songs and calls 34-7
  tail shapes 63
  territory 40-2
  wings **18-19**, 20, **21**, 63
blackbirds **11**, **41**, 42
  Brewers's **13**
  red-winged **13**, **99**, 114, 124
  yellowheaded **35**
blackcaps **149**
blackcocks 42
bluebirds 113
  Eastern 63
  mountain 139
bobolinks 114, 149
bobwhite 121
  northern 115, 117
bogs 94-5
bramblings **10**
breebirds, Eastern **16**
breeding season 28
British Royal Society for the Protection of
  Birds 152, 154
British Trust for Ornithology 142, 143
brood patch 50
brush 116-17
buntings
  lark 115
  painted 117
  snow 115
bustards 153
  great 34, **42**
buzzards
  roughlegged **63**, **75**

C

cameras 69
capercaillies 42, **43**
cardinals 10, **10**, 12, 14, 79, 113, **117**
cassowaries 8
catbirds 113, 117
  gray **116**
censuses 142-5
chacchalacas 125
chickadees 12, 16, 117
  black-capped 14, **123**

boreal 121
Carolina 14
mountain **138**, 139
chiffchaffs 85
chuck-will's-widows 99, 121
cliff colonies 100-1
Common Bird Census 143, 144
condors, California 55, **152**
conservation 152-5
coots **27**, **90**
cormorants 53
  double-crested 92, 109
  flightless 18, 19
cowbirds 49
cranes 152
  sandhill 97, 132
  whooping 97, 155, **155**
creepers
  brown 38, **123**
  tree 38
crossbills 24, **28**, 120
  red 121
  white-winged 121, **121**, 133
crows 36, 79, 81, 83, 113
  fish 105
cuckoos 46, 49, **49**
  black-billed 116
  Eurasian 51
  European 116
  mangrove **124**, 125
  yellow-billed 116, **116**, 125
curlews 24, **26**
  long-billed 105, **105**

D

desert and semi-desert 126-31
dickcissels 115
dippers, North American 86, **86**
display 42-3, **42-3**
doves 46, 53
  mourning 12, 13, 14, **78**, 79
  rock 46, 76
  white-fronted 125
  white-winged 129
dowitchers 105
ducks 23, 24, **27**, 53, 54, 75
  American wigeon 96, 109
  black-bellied whistling 93
  blue-winged teal 96, **98**
  bufflehead 89, 91, 109
  canvasback 91, 96, 109
  cinnamon teal 92
  eider 110
  common 111, 132
  king 111, 132, **133**
  spectacled 132
  Stellar's 132
  fulvous whistling 93
  goldeneye 89, **90**

Barrow's 109
common 109
goosander **89**
green-winged teal 96, **98**
harlequin 86-7, **87**, 111
long-tailed **110**
mallard 53, 55, 91, **91**, 96, 110,
  mergansers
  common 89
  hooded 109
  red-backed 24, **24**
  red-breasted **107**, 109
  oldsquaw 111
  pintail **90**, 92, 96
  puddle 132
  redhead 91, 92, 96, 109
  ring-necked 91, 109
  ruddy 96, 109
  scaup 91, 96, 109
  scoter 111
    black 111
    surf 111, **111**
    white-winged 111
  shoveler **90**, 92, 96
  wood 89, 93
dunes and rock beaches 102-3
dunlins 37, 105

E

eagles 24, **27**, 51, 53, 63
  bald **24**, 26, 75, 91, **144**
  golden 9, **9**, 26, 40-1, 75, 134,
    134, 136
eggs 48-9, **48-9**
  hatching 52-3
  incubation 50-3
egrets 98
  cattle **80**, 93
  great 93, 107
  reddish 107
  snowy 93, 107
elephantbird 48
estuaries 106-9
ethology 140-1

F

falcons 26
  gyrfalcon 133
  peregrine **21**, 48, **63**, **76**, 77, 133,
    143, **144**, 153, **153**
  prairie 131
  red-footed 45-6
farms 112-13
feathers 22-3, **22-3**
feeding birds 11, 12-15, **13-15**
field guides 64-5, **65**
field notebooks 66-7, **66**

finches **12**, 45, 75
  American goldfinch 10, 74
  Cassin's 139
  chaffinch **10**
  goldfinch 14
  greenfinch **10**
  hawfinch **10**
  house finch **11**, 14, **78**, 79
  purple 121
  rosy 86, 133, 135, **135**, 136, 137
flamingos 26
  Andean 39, **39**
  Chilean **38**, 39
  greater 31, **38**, 39
  James's 39, **39**
  lesser 39, **39**
flight 8-9, 18-23, **18-23**
flycatchers 84
  alder 95
  Arcadian 119
  brown-crested 129
  great crested **67**, 119, **123**
  greater kiskadee 125
  Hammond's 139
  least 119
  olive-sided 94, 139
  pied **68**, 85
  spotted **67**, 85
  vermilion **28**
  western 139
  willow 117
  yellow-bellied 94, 95, **95**
forests 118-25
fulmars 48, **55**, 101, **140**

G

gannets 19, 41, **41**, 101, 143
geese 23, 53, 75
  barnacle 145, **145**
  brant 109, 132
  brent 55, **108**
  Canada 75, 91, 99, 109, 132
  emperor 132
  greylag **141**
  Hawaiian 152
  Richardson's Canada 91
  Ross's **31**, 132
  snow, **31**, 33, 75, 109, 132
  white fronted 132, 148
gnatcatchers 129
godwits 24, 63, **109**
goshawks 75, 120, 121, 138
  northern **121**
grackles 14, 77, 79, 81, 83, 113
  boat-tailed 105
grasslands 114-15
grebes 24, 43, 47, 92
  eared 92
  great crested **41**, 43

horned 92, **93**
  pied-billed 92
  red-necked 92
  Slavonian **93**
  western 92
grosbeaks 13, 14, **15**, 24
  blue 113
  evening 75, 121
  pine 121
grouse 10, 19, 41, 42, 50, 52, 55, 121
  blue 138, **139**
  red 51
  ruffed **34**, 36, 42, 118, 121, 138
  sage 42, **42**, 130, **130**
  sharp-tailed 130
  spruce 120, **120**, 121, 138
  Franklin's 138
guillemots **100**, **101**
  black 101
  pigeon 100
  small black 100
gulls **80**, 83
  California 92
  Franklin's 92
  glaucous-winged 81, 102
  great black-backed 81, 101, 102
  herring 80-1, 89, 101, 102, **103**
  laughing 102, 103
  ring-billed 81, 83
  Ross's **73**
  western 81, 102

H

hacking 77
harriers 63
  northern 109, 115, **115**
hatching of eggs 52-3
hawks
  broad-winged **46**, 75
  Cooper's 15, 77
  ferruginous 115, 131, 148
  Harris's 126, **127**
  red-shouldered 99, **122**
  red-tailed 75, 122, 131
  roughlegged **63**, 133
  sharp-shinned 15, 77, 121, 122, 138
  Swainson's 75, 115, 131
herons 24, 98
  black-crowned night-heron 93
  great blue 83, 89, 92, **96**
  green-backed 89, 93, **93**
  little blue 93, 107
  tricolored 107
  yellow-crowned night-heron 93
hummingbirds 20, 26, 48
  broad-tailed 139
  calliope **47**, **136**, 137, 139
  Costa's 127
  rufous 139

sword-billed 27

I

ibises, white **80**, 98
identification of birds 62-3
incubation and rearing young 50-3
intertidal zone 104-5

J

jacanas 24, **27**
jaegers
  long-tailed 133
  parasitic 133
  pomarine 133
jays 12, 121
  blue 14, **14**, 79, 119, **123**, 138
  gray 120, 138
  green 125
  Stellar's 138
juncos 10, 12, 13, 14, 75
  dark-eyed 118
  Oregon 10

K

kestrels 16, **22**, **112**
  American 113, 122
killdeers **48**, **50**, 115
kingbirds
  Couch's 125
  gray 125
kingfishers 44
  belted 88, **89**, 125
  green 125
  ringed 125
kinglets 48
  golden-crowned 94
kites, snail 27
kittiwakes 37
  black-legged 101
kiwis 48
knots **74**, 107, **107**, 148
  red 103, 105

L

larks, horned 105, 131
limpkins 99
longspurs
  chestnut-collared 115
  Lapland 115, 133, **133**
  McCowan's 115
  Smith's 133
loons 90-1, **92**

common 132
  red-throated 132, **132**
  yellow-billed 132
lyrebirds 36

M

martins 25
  house 17, **150**
  purple **16**, 17
  sand 17
meadowlarks
  eastern **112**, 113, 114
  western 115
merlins 121
migration 30-3, 74-5, 102-3
  studies 146-51
mimicry 36
mockingbirds 14, **17**, 36, 77, 79, 113
molt 23
moorhens **89**
  common 88
mountain streams 86-7
mountains 134-5
murrelets 100
  marbled 100
murres 48, 101
  common 100
  thick-billed 100

N

nest boxes 16-17, **17**, 89
nests 44-7, **44-7**
nighthawks 35, 37, 44
  common 37, 76-7, **77**
  lesser 126
nightingales 35, **35**
nightjars 35, 37, 44
  pennant-winged 21
northern flickers 79
nutcrackers, Clark's 138, **138**
nuthatches 16
  brown-headed 121
  red-breasted 139
  white-breasted 14, **14**

O

oceans 110-11
orioles
  Altimira 125
  golden **69**
  northern **45**, 88, 113
  orchard 113
ospreys 24, 28, 54, 75, 91
ostriches **8**, 18, 22, 24, 34, 48

ovenbirds 118
overshooting 74
owls 16, **18**, 35, 50, 51, 53, 75, 129, 152
  barn **19**, 113, **113**
  barred 98, 122
  boreal 121
  elf 128-9, **129**
  great gray 37, 121, 138-9
  great horned 28, 77, 81, **81**, 121, 122, 138
  long-eared 37
  northern hawk-owl 121
  screech 77
    eastern **119**, 122
  short-eared 109, **114**, 115
  snowy **74**, 115, 133, 138
oystercatchers **51**, **105**
  American 105
  black 103

P

parrots **55**, 129
peacocks 42
peeps **62**
pelicans 53
  white 92
penguins 8, 19, 22
  emperor 8, **8**
petrels 110
  Leach's 101
  storm 44, 110, 111
  Wilson's **111**
pewees, eastern wood 119
phainopeplas 129
phalaropes 50
  red 111
  red-necked 111
pheasants 48, 113, 152
  ring-necked 113, 114, 115, **115**
phoebes 16
photography 68-9
pigeons 46, 53
  red-bellied 125
  white-crowned 124-5, **124**
  woodpigeon **52**
pipits 25
  meadow **49**
  water 137
pishing 117
plovers 45, 53, 74
  black-bellied 49, 83
  gray **109**
  lesser golden 83
  mountain 115, 127
  piping 102
  ringed 42-3, **45**
  semipalmated **26**, 42-3
plumage 20-1, 22-3, 42-3

ponds and lakes 90-3
poor-wills 30
populations 144-5
potholes 96
prairie chickens 35
  greater **114**, 115
  lesser 115
ptarmigans 25
  rock 133
  white-tailed 136-7, **137**
  willow **63**, 133
puffins 86, **100**, 101, 110
  Atlantic 100-1, **100**
  horned 100
  tufted 100

Q

quails
  California 126
  Gambel's 126, **126**
  mountain 126
  scaled 126
quetzals 21

R

rails, 18, 107
  black 108
  clapper **106**, 107
  king 88, 99
  sora 108
  Virginia 108
rarities 72, 75
ravens **36**, 121, 135, **135**, 136
razorbills 100-1, **101**
recording birds 70-1
red shanks, spotted **75**
redpolls 13
  common 133
  hoary 133
reed buntings **49**
resource partitioning 105
reverse sexual dimorphism 88
rheas 8
rivers 88-9
roadrunners 129, **129**
robins 16, **40**, **44**, **50**, **79**, 113, 142, **148**
  American 10, **54**, 55, **55**, 77, 78
  Eurasian **141**
  European 10, **43**, **55**, 140, 149
rooks 45-6

S

sagebrush flats 130-1
sanderlings 37, 104, **105**, 133, **148**

gray 29
sandpipers 74, 103
  common 87
  least 105
  pectoral **83**, 133
  purple **104**, 105
  rock 133
  semipalmated 103
  spotted **86**, 87, 89
  upland 114, 115
shearwaters 110
  Cory's 111
  flesh-footed 111
  greater 111
  Manx **110**, 111, **147**
  short-tailed 111
  sooty 111
shrikes, northern 133
siskins, pine 12, 121
skimmers
  black 27, 107, **108**
skuas 21
snipes **26**
solitaires, Townsend's 139
songs and calls 34-7
sparrows 13
  American tree 14, 115, 133
  Bachman's 121
  Baird's 115
  black-throated 129
  Brewer's 131
  Cassin's 115, 127
  chipping 79, 113
  fox 133
  grasshopper 114
  Henslow's 114
  house 76, 83, 113
  lark 115
  Le Conte 115
  Lincoln's 94, 139
  North American 10
  olive 125
  pine-woods 121
  sage 131
  savannah 105, 114
  seaside 105, 109
  sharp-tailed 109
  song 14, 35, 79, 113, **140**
  vesper 114
  white-crowned 137, 139
  white-throated 75, 120, 121
spoonbills **12**, 24
  roseate 125, **125**
starlings 14, 24, **35**, 36, 55, 76, 78, 81, 83, 113
stilts, black-necked 25, **83**
stints 62
storks 30, 150
  white 28, **31**, **150**
  wood 98-9
surfbirds 103

swallows 25, 32, **63**, 147
  bank 17, 47
  barn 17, **32**, 47, **63**, 112-13, **112**, 147
  cliff 92
  tree 32, 75, 88, 92, 105
  violet-green 92
swans
  mute **90**
  trumpeter **96**, 97-8, **147**
  tundra 109, 132
swamps and marshes 96-9
swifts 19, 20, 25, **27**, **55**
  black 77
  chimney 77
  tree 32
  Vaux's 77
  white-throated 77

T

tanager
  scarlet **118**, 119, 123
  summer 123
  western 139
telescopes 58-9, **58-9**, 60, **61**
terns 81, **102**
  Arctic **20**, 30, **30**
  black 97, **97**
  Caspian 107
  common 97, 106, 154
  elegant 107
  Forster's 92, 97, 106
  gull-billed 107
  least 45, **102**
  roseate 81, 102, **103**
  sandwich 107, 154
  sooty **57**
territory 40-3
thrashers
  Bendire's 127
  brown 113, 117
  curve-billed 127, **127**
  sage 131, **131**
thrushes 17, **37**, 48, 84
  hermit **85**, 139
  mistle **53**, **69**
  song **37**, **44**
  Swainson's 139
  wood 116, 118, **118**
titmice 12, 16, 25
  tufted 15, 117
tits
  blue **10**, 52
  great **10**, 34, **51**
  long-tailed **47**
  willow 16
towhees
  rufous-sided 118, **123**

INDEX

turkeys, wild 42, 118, 119, 121
turnstones 105, 107, **109**
  black 132-3
  ruddy 103

V

verdins 129
vireos 84
  blackwhiskered 124
  red-eyed 116, 191, **123**, 124
  solitary 119
  warbling 88, 139
  white-eyed 117
  yellow-throated 119
vultures 20
  black 81, **81**

W

wagtails 25
  yellow 133
warblers **12**, 13, 84
  arctic 133
  black-andwhite 119, 123

black-throated blue 119
black-throated green 123
blackburnian 123
blue-winged 119
Cape May **120**, 121
cerulean 119
chestnut-sided 117
garden **52**
golden-winged 117
hooded 118
icterine **73**
Kentucky 118
MacGillivray's 139
magnolia 121
melodious **73**
Nashville 94
orange-crowned 133, 139
palm **94**, 95
parula 99, 123
pine 121
prothonotary 88, 99
sedge **49**
willow 85
Wilson's 133, 139
wood 85
worm-eating 118
yellow **84**, 113

yellow-throated 88, 99, 117, 119
waterthrushes
  Louisanna 87
  northern 87
waxwings
  Bohemian 133
  cedar 74, 113
weavers 65
welands 96-7
wheatears **151**
  northern 133
whimbrels **23**
whip-poor-wills 118, **122**
whitethroats 85, 145
  lesser 85
Wildfowl Trust 152
woodcocks **49**, 55
woodpeckers 12, 24, **27**, 35, 44, 129
  downy 14, 117, **123**
  gila **128**
  hairy **34**, 122
  ivory-billed 122
  pileated 99, 122-3, **123**
  red-cockaded 121
  three-toed 121
wrens 9, **9**, 16, 44, **49**
  Bewick's 117

cactus **128**, 129
Carolina 117
house 113
rock 126
winter 88, 94, **94**

Y

yellowlegs **26**
  greater **82**

ACKNOWLEDGMENTS

Quarto would like to thank the following for providing photographs
and for permission to reproduce copyright material:

37 (left) National Sound Archive/British Library of Wildlife
Sounds/Richard Ranst 70 ARDEA 71 Jack Skill.
All other photographs supplied by Bruce Coleman.

Artwork provided by:
Peter Bull
David Kemp
Mark Iley
Janos Marffy
Maurice Pledger
Paul Richardson
David Thelwell